# Parker's Illustrated Home Guide to Plumbing

HERBERT T. LEAVY

PARKER PUBLISHING COMPANY, INC.
West Nyack, N.Y.

Library of Congress Cataloging in Publication Data

Leavy, Herbert T
    Parker's illustrated home guide to plumbing.

    Includes index.
    1.  Plumbing--Amatuers' manuals.  I.  Title.
TH6124.L4          644'.6          77-20116
ISBN 0-13-650275-X

Printed in the United States of America

*This book is dedicated to
my wife and friend, Patricia*

# HOW THIS BOOK WILL SAVE YOU
# TIME AND MONEY

This heavily illustrated, practical guide to plumbing was designed with one purpose in mind—to show you how to save time and money by doing your own plumbing maintenance, repairs and replacement. The book is just as valuable to the person building or buying a new home as it is to the apartment dweller and owner of an older home, since you are guided step-by-step through emergency situations, the selection of pipe, fixtures and fittings . . . to the installation, maintenance and eventual repairs.

A first quick reading of the book will give you an understanding of your entire plumbing system, and allow you to comprehend each component of the system in much clearer fashion. Then, reference to those sections which most directly affect what you are planning to do (an emergency repair or installation of a new sink, for example) will be clearer to you as your plumbing projects take shape.

Most plumbing books on the market today approach each piece of work to be done as an entity in itself, with little reference to the system as a whole. Since a plumbing system is so inter-related—the water supply, fixtures, the drainage system—we have taken the broader view so that when you do a job, you will know exactly how it can affect the rest of the plumbing. This over-all approach is easy to master, and in the long run can save on costly additional repairs or replacements.

For example, before you install plumbing fixtures we explain how it can affect your drainage system. Overloading. your present drainage system could lead to back-up problems. Recognizing this beforehand is the sensible approach, and we show you how to anticipate this and other potential problems.

Once you are acquainted with the total system, we show you the relatively few, correct tools to use, and how to use them. The investment of a small number of dollars in quality tools and the knowledge of how to use them properly can save you hundreds, even thousands of dollars in plumbing bills. The cost of this book will probably be recovered by the savings you will enjoy by doing your first small plumbing repair or installation.

Concerning tools, more money has been wasted by the amateur plumber who uses a pair of pliers when he should be using a wrench, or by using a too-small screwdriver, than has ignorance of the job at hand. The wrong tool can damage expensive fixtures, creating frustation and a bad-looking or non-functioning job. This book shows you how to avoid such problems before they occur.

What is true for tools is equally true for pipe and fittings. There is a job which each kind of pipe and fitting is intended for, and it is not difficult to learn through our heavily-illustrated plumbing-teaching system.

The fact is that the most expensive room on a per-square-foot basis is the bathroom. Fixtures and plumbing are expensive, and should be selected, installed and maintained with care. And, equally important, the handyman should be instructed on how and when to call in a professional plumber. We offer the guidelines which can keep you out of trouble and in a satisfied frame of mind—satisfied that you have achieved your goal of a smoothly running plumbing system at a minimum cost in time and money.

Coping with plumbing emergencies can—really—be a satisfying experience if you are well prepared. We show you how to maintain your peace of mind and your plumbing at the same time. Having an emergency tool and parts kit on hand, for example, complete with the things you will probably need, is a start. A sensible precaution, but too few take it. We show how to stock the kit, and lead you through all of the emergencies which are likely to arise. And we do it in such clear cut fashion that you'll find yourself wanting to instruct the rest of the family in these emergency repairs—in itself, another commendable idea, since anyone can be home alone when an emergency strikes.

Preventive maintenance is the key to a happy household in terms of the plumbing system, and we show you exactly how to go about setting up a program that will eliminate the type of emergency you now have or anticipate having. For example, when you replace that washer in the hot water faucet, do you at the same time replace the one in the cold water faucet, even though it's not leaking? You should. It only takes another minute, you have the water shut off, the tools and equipment at hand, and once done you are covered in *that* sink or bathtub for a longer period of time. Are you feeding your cesspool or septic system the right quantity of chemicals on a regular basis, if needed? If not, you may be asking for expensive trouble.

We show you, in this book, the easy way to avoid problems *before* they start. Once you embark on a program of knowing your system, checking it regularly, doing the correct things to insure high performance, you will be doing *less* plumbing than your neighbors who wait for a crisis to happen.

Let's face it, plumbing is not a hobby. It's a chore, and a certain amount of work. But knowing the professional way to do things—which we show you—will keep those chores at a minimum, and save you time and dollars.

**Herbert T. Leavy**

**Acknowledgments**: We wish to thank the following individuals and companies for their help in freely furnishing us with advice, the latest technical information and material which was of great assistance in the preparation of this book. Thanks to ITT Grinnell Corporation, 260 W. Exchange St., Providence, R.I. for permission to use selected material from their *Pipe Fitters Handbook*. Thanks to Daniel W. Irvin, Educational Director for Rockwell International, for permission to use special material, and thanks to the Montgomery Ward Public Relations department for their most helpful cooperation in furnishing technical help and material.

# CONTENTS

# 1

# HOW TO MAKE EMERGENCY PLUMBING REPAIRS

It might happen on one of your busier days—a holiday when the house is full of guests—or it might occur in the middle of the night: The plumbing emergency, when the best thing to do is not panic, but be totally prepared.

The commode could begin to overflow and no amount of jiggling the control handle can stop the water from edging toward wall-to-wall carpeting. Or it could be the sudden bursting of a pipe, an out of control faucet or an overflowing washing machine.

Such emergencies are inconvenient and can cause damage running into thousands of dollars, if corrective action is not taken immediately. Being prepared for plumbing emergencies is a simple thing, but rarely understood in most homes. A few minutes of your time now could be insurance for future security, and trouble-free performance of your water system.

## THE EMERGENCY PLUMBING KIT

The basic elements in your emergency "preparedness kit" (see Figure 1–1) include "O" rings in an assortment of sizes, faucet washers with screws in assorted sizes, compression nuts and rings, replacement faucet seats, liquid and dry type drain cleaners, a faucet seat dresser, a propane torch, solder and flux, 1¼" and 1½" trap washers, graphite packing, a plumber's force type plunger (also known as a "plumber's helper"), a tube cutter, a flaring tool, some electrician's tape, and extra elbows,

**Figure 1-1**
**It's a good idea to assemble**
**your own first aid plumbing kit.**

sleeves and nipples. Also—just in case the emergency will turn out to be more than you can cope with, have the number of a reliable plumber offering round-the-clock service. Basic tools and other materials include a medium sized pipe wrench and adjustable end wrenches, assorted screwdrivers including Phillips head, adjustable pliers, pipe joint compound, and plumber's putty.

## CLOGS AND HOW TO DEAL WITH THEM

Drains may become clogged by objects dropped into them or by accumulated grease and other matter. If something valuable, say a ring, is dropped into the kitchen sink and can't be retrieved, the fixture trap can be easily removed. Don't run water in this event, or try anything else. Below the sink drain, you will find a large, curved pipe called a "fixture trap." The drained water runs into this trap and objects are literally "trapped" by the curve in the pipe. Put a pan below the pipe and loosen the two large nuts at either end (some traps have a clean out at the bottom which can be removed, so you don't have to take off the pipe). Keep turning and the whole curved piece of pipe will drop off in your hands. Note the washers used—they go back when you reinstall the fixture trap after cleaning.

*With a clogged sink drain* that is not completely clogged, try scalding water for five or ten minutes. If that doesn't do the trick, use a strong chemical drain cleaner, carefully following label directions. If the chemical opens the drain, be sure to flush with hot water for at least ten minutes.

If you have a complete clog and water in the sink, don't use a chemical cleaner. Chemicals must be in direct contact with the stoppage and should be used only for minor clogs, and handled with extreme care, since most contain lye, which is deadly. First try a suction cup force plunger, removing basket strainer from the drain and making sure there's enough water in the sink to provide a seal. Cup the plunger tightly over the drain and pump the wooden handly vigorously several times. If this doesn't work, the next step is to remove the cleanout plug if the trap has one. If you still have an obstruction, use a snake or drain auger. See Figure 1–2 for illustration of these various steps. For best results with a snake or drain auger, feed it in and then rotate, repeating until the obstruction is clear. Once clear, use scalding water for ten minutes to carry away grease or other accumulations.

*A clogged lavatory or tub* may have a pop-up stopper. The most likely cause of stoppage is an accumulation of hair, etc., in this mechanism. See Figure 1–3. To remove one type of lavatory stopper, unscrew the ball joint (at back of the drain under the basin) and disengage the lever from the stopper by pulling it straight back. The stopper can now be lifted out for cleaning. Be sure to retighten the ball joint properly when replacing it. To remove a tub stopper (or similarly made lavatory stopper) simply rotate it 90-180° and lift it out.

You can use a length of wire with a hook on it to pull out any debris that may lie under the drain stopper. If this doesn't work, follow the procedure described under sink drain. A lavatory or tub will have an

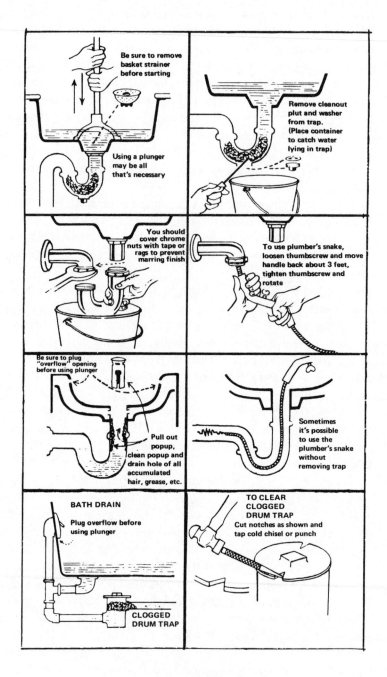

**Figure 1-2**

**How to clean out clogged drains.**

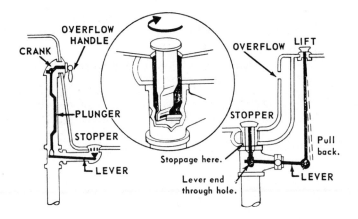

**TYPICAL TUB AND LAVATORY STOPPERS**

**Figure 1-3**

overflow outlet, which should be closed with rags or tape when using a force plunger.

A tub has a drum trap—which may be behind the tub (reached by an access panel in the partition behind the tub), in the basement under the tub, or embedded in the bathroom floor. The entire cover screws out of the trap body. A fine oil applied all around the cover edge will help to loosen "frozen" threads . . . or you may have to use a cold chisel and hammer to free the cover (see Figure 1–4). When replacing it use a new gasket if the old one is damaged and apply cup grease to the threads and gasket to obtain an easier, better seal.

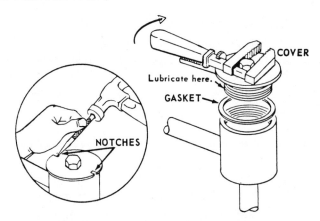

**REMOVING A DRUM-TRAP COVER**

**Figure 1-4**

If the preceding suggestions fail to remedy the trouble, stoppage is most likely to be in the branch drain. When a little water will seep through, continued use of scalding water and/or one of the commercial preparations available for opening clogged drains may, in time, open the drain. To use a commercial preparation, follow directions and unless specifically told otherwise, pour it directly into the drain (with stopper out) through a funnel, to avoid damaging the fixture finish. (See Figure 1–5)

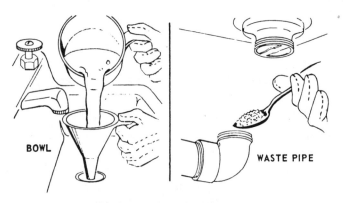

BOWL

WASTE PIPE

**TWO WAYS OF USING A COMMERCIAL DRAIN SOLVENT**

**Figure 1-5**

Should all else fail, a cleanout auger will have to be used (see Figure 1–2).

*A floor or areaway drain stoppage* is usually caused by an accumulation of lint, silt or leaves in the trap—or, possibly, in the horizontal portion of the run. Use of a small auger (after removing the strainer cover) will clear the trap; but a large auger or an electric cable will be required (see Figure 1–6) if stoppage is in the line. In the case of a gutter drain, either of the latter will have to applied, starting at the top of the downspout—or downspout will have to be disconnected at the bottom for cleaning of the drain (the obstruction will generally be lodged in the bend at the foot of the downspout). The use of gutter leaf guards will prevent this type of trouble.

If the use of house plumbing causes a backup of water out of a basement floor drain (or any lowest drain in the system), either the house drain or the house sewer is stopped. These generally can be easily cleaned by use of an electric cable (they can be rented) fed in through the foot-of-

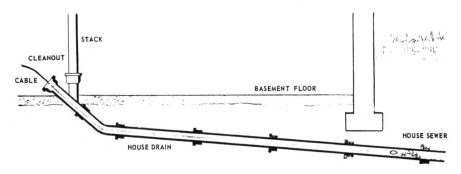

**Figure 1-6**

the-stack cleanout (see Figure 1–6) provided for this purpose. However, if roots have penetrated into the house sewer to cause the problem, a special root-cutting cable is required (now you are in the professional plumber's area), and relief will be only temporary unless the defective pipe is replaced with root-proof pipe and joints. Any backup caused by a heavy downpour indicates a public sewer problem to be corrected by the sewer authorities.

A facility that is seldom used (basement floor drain, toilet, etc.) may lose its trap water through evaporation—and allow sewer gas to escape. Occasional filling of the trap through use will remedy this. Also, kerosene (or any oil) floated on the water will slow its evaporation. If these remedies fail the system is improperly vented and needs redesigning.

*If you have a clogged toilet,* first try the force plunger and be sure to have sufficient water in the bowl during pumping. If this fails (see Figure 1–7) use a closet auger, which is a short snake with a crank on one end and a hook on the other. Start the snake into the bowl and crank until it becomes tight, then pull it back, an action that will often bring up the obstruction. If this doesn't work, use a small snake in the same way as described in opening drains. The final resort: the toilet must be removed from the floor. Don't panic. You can do it.

*If an obstruction in the water closet trap or leakage around the bottom of the water-closet bowl requires removal of the bowl,* follow this procedure:

Shut off the water. Empty the tank and bowl by siphoning or sponging out the water. Disconnect the water pipes in the tank (see Figure 1–8). Disconnect the tank from the bowl if the water closet is a two-piece unit. Set the tank where it cannot be damaged. Handle the tank and bowl carefully since they are made of vitreous china or percelain and are easily chipped or broken.

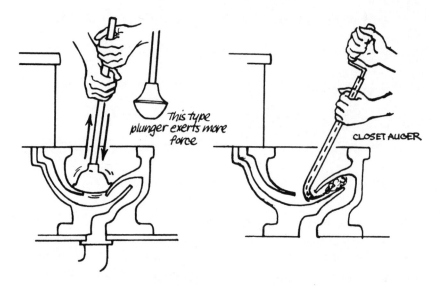

This type plunger exerts more force

CLOSET AUGER

**Figure 1-7**

**Cleaning out a clogged toilet.**

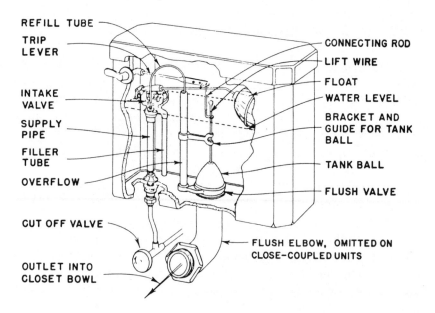

REFILL TUBE

TRIP LEVER

INTAKE VALVE

SUPPLY PIPE

FILLER TUBE

OVERFLOW

CUT OFF VALVE

OUTLET INTO CLOSET BOWL

CONNECTING ROD

LIFT WIRE

FLOAT

WATER LEVEL

BRACKET AND GUIDE FOR TANK BALL

TANK BALL

FLUSH VALVE

FLUSH ELBOW, OMITTED ON CLOSE-COUPLED UNITS

**Figure 1-8**

Remove the seat and cover from the bowl. Carefully pry loose the bolt covers and remove the bolts holding the bowl to the floor flange (see Figure 1–9). Jar the bowl enough to break the seal at the bottom. Set the bowl upside down on something that will not chip or break it. Remove the obstruction from the discharge opening. Place a new wax seal around the bowl horn and press it into place. A wax seal (or gasket) may be obtained from hardware or plumbing-supply stores.

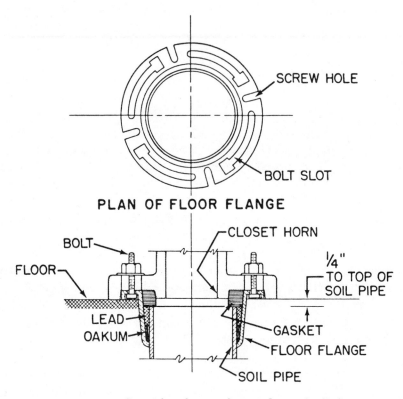

**PLAN OF FLOOR FLANGE**

Connection of water closet to floor and soil pipe.

**Figure 1-9**

Set the bowl in place and press it down firmly. Install the bolts that hold it to the floor flange. Draw the blots up snugly, but not too tight because the bowl may break. The bowl must be level. Keep a carpenter's level on it while drawing up the bolts. If the house has settled, leaving the floor sloping, it may be necessary to use shims to make the bowl set level. Replace the bolt covers.

Install the tank and connect the water pipes to it. It is advisable to replace all gaskets, after first cleaning the surfaces thoroughly. Test for leaks by flushing a few times. Install the seat and cover.

*Caution*: augers, rubber force cups and other tools used in direct contact with sewage are subject to contamination. Do not later use them to work on your potable water supply system unless they have been properly sterilized.

Always use the proper size wrench or screwdriver. Do not use pipe wrenches on nuts with flat surfaces; use an adjustable or open-end wrench. Do not use pipe wrenches on polished-surface tubings or fittings, such as found on plumbing fixtures; use a strap wrench. Tight nuts or fittings can sometimes be loosened by tapping lightly with a hammer or mallet.

### LEAKING FAUCETS

Faucets in the past all had washers to control the flow of water. If a faucet drips when closed or vibrates, ''sings,'' or ''flutters'' when opened, the trouble is usually the washer at the lower end of the spindle. If it leaks around the spindle or shaft when opened, new packing is needed. To replace the washer, first cut off the water at the shutoff valve nearest the faucet.

Take the faucet apart (see Figure 1–10)—the handle, packing nut, packing and the spindle, in that order. You may have to set the handle back on the spindle and use it to unscrew and remove the spindle.

Remove the screw and the worn washer from the spindle. Scrape all worn washer parts from the cup and install a new washer of the proper size and angle. For just the right part, take the old washer to a hardware or other store for matching.

Examine the seat on the faucet body. The washer fits the seat and if it is rough or badly worn around, a new washer won't help the leaking faucet. Your local stores carry a seat dressing tool for refacing faucet seats. Reassemble the faucet—handles of mixing faucets should be in matched positions.

To replace the packing, simply remove the handle, packing nut, and old packing, and install a new packing washer. If a washer is not available, you can wrap stranded graphite-asbestos around the spindle. Turn the packing nut down tightly against the wicking.

Many modern water controls combine convenient single-handle operation with a uniquely designed cartridge in place of washers. The self-adjusting cartridge is simply dropped into place. The result is a brand-new

## EMERGENCY REPAIRS – LEAKING FAUCETS

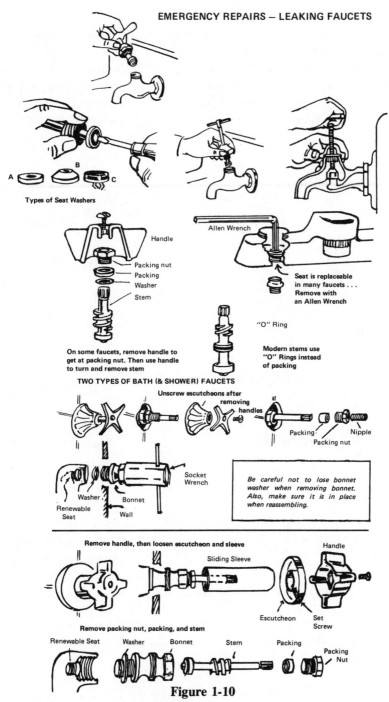

Types of Seat Washers

Handle

Packing nut
Packing
Washer
Stem

On some faucets, remove handle to
get at packing nut. Then use handle
to turn and remove stem

Allen Wrench

Seat is replaceable
in many faucets . . .
Remove with
an Allen Wrench

"O" Ring

Modern stems use
"O" Rings instead
of packing

### TWO TYPES OF BATH (& SHOWER) FAUCETS

Unscrew escutcheons after
removing handles

Packing          Nipple
Packing nut

Socket
Wrench

Washer
Renewable          Bonnet
Seat          Wall

Be careful not to lose bonnet
washer when removing bonnet.
Also, make sure it is in place
when reassembling.

Remove handle, then loosen escutcheon and sleeve

Sliding Sleeve          Handle

Escutcheon          Set
Screw

Remove packing nut, packing, and stem

Renewable Seat          Washer          Bonnet          Stem          Packing          Packing
Nut

### Figure 1-10
### Leaking faucets repair.

faucet because all wear takes place in the cartridge itself, thus eliminating the chore of refacing the seat required with ordinary faucets. Replacing a cartridge takes only a screwdriver, or in some models no tools at all.

To replace the washers in some types of bath faucets, you will need an extended socket wrench of the right size. First, remove the handle and escutcheon (see Figure 1–10). If there happens to be a sleeve over the stem, protect it with tape and then unscrew it. Fit the socket wrench over

## EMERGENCY REPAIRS — LEAKING TOILET

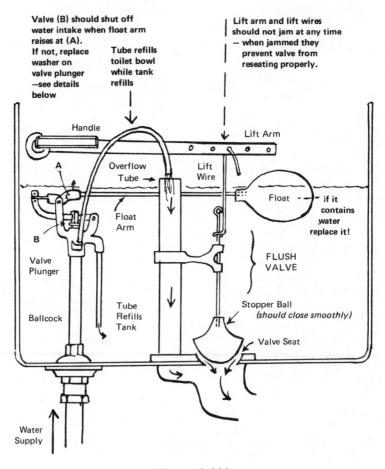

**Figure 1-11A**

the bonnet and remove—the bonnet and stem will come out of the body together. Replace the washer and reassemble.

### THE LEAKING TOILET

There are three basic types of flush-tank disorders, and there is a different cause for each. The cause and cure of each is not overly complicated.

*If the tank fills, but the water still runs,* the chances are you have a defective ballcock (see Figure 1–11). Check for a worn washer on the stem for the problem, but first test by gently lifting on the float arm. If the water shuts off, repair is easy—just bend the float arm until the float is about half an inch lower, or until the tank fills about one inch from the top of the overflow pipe. If you have an old metal style float, check to see that it isn't waterlogged through a leak and cannot rise high enough to shut off the water; if the water has leaked into the float ball, replace it.

*If the water does not shut off when you raise up on the float arm,* then the trouble is in the ballcock valve itself. To replace the washer, first shut off the water and flush the tank to clear it. Remove the two thumbscrews (Figure 1–11B) that hold the float assembly onto the

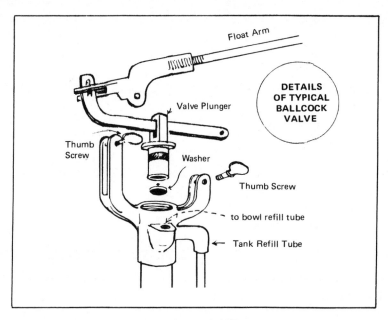

**Figure 1-11B**

**Leaking toilet repair.**

ballcock valve. Slide the plunger and attached linkage out of the valve. It is then easy to remove the washer with another of the same type. If your ballcock has seen very long service, it might be wiser to replace it than to attempt a repair.

*If the water runs but the tank does not fill*, you may have a problem caused by a worn or hardened flush ball (Figure 1–11A), the stopper that controls the flow of water from the tank to the bowl. Before replacing it, however, check first to see that the lift wire moves up and down in its guides without binding. If the action is free and the ball seats squarely on the flush valve seat, then the rubber flush ball needs replacing. To do this, hold the ball in one hand and unscrew the lower lift wire with the other. If the wire sticks in the ball, use pliers to remove it. Before replacing the flush ball, clean the valve seat with steel wool. If the flush ball does not drop directly into the seat, loosen the setscrew on the guide and adjust the guide so that the ball drops directly into the correct position. Then re-tighten the guide screw.

*If the flush tank handle must be held down constantly while flushing tank*, this is no doubt a disorder caused when the flush ball is not lifted high enough out of its seat during flushing. Suction caused by the water rushing through the seat pulls the flush ball back into its seat prematurely, stopping the flushing operation unless the handle is held down. This disorder is remedied (see Figure 1–11A) merely by shortening the linkage wire raising the ball.

## LOCATING AND FIXING LEAKING VALVES

When your plumbing system was installed, the contractor usually anticipated possible emergencies by placing a series of shutoff valves throughout the home. We suggest you at some point tour your home with a flashlight, an adjustable wrench, and a set of colored tags you can key to label every control.

*The main shutoff valve* is at the water line coming into the home from the street. It is immediately connected to the water meter followed by a master shutoff valve for the entire home. Close this one valve (see Figure 1–12) and you have shut off water throughout the entire home—it's instant action for serious emergencies, and for repairing fixtures which don't have their own shutoff valve.

*The kitchen shutoff valve* is below the kitchen sink—you'll find both hot and cold water valves. In some cases, the valves will be below the kitchen in the basement. Craftsmen plumbers sometimes install two sets—at the sink and in the basement.

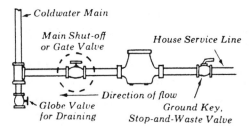

*Coldwater Main*

*Main Shut-off or Gate Valve*

*House Service Line*

*Direction of flow*

*Globe Valve for Draining*

*Ground Key, Stop-and-Waste Valve*

**Figure 1-12**

*Bathroom valves* are in the lavatory, commode and bathtub. They are usually below the fixture for easy access. Tub valves can be in the basement or behind an access plate in back of the faucet controls. The commode or water closet has a single cold water valve normally installed below the water tank.

If you have a hot water heating system, you will find a cold water shut-off valve near the furnace. Each radiator also has individual shutoff valves on one end of the unit.

Just about all that's left, serviced by water, are your water-using appliances. These, you will find, have shutoff valves located on or near the appliance.

You are ready for any water emergency when you know where the valves are located for all your faucets and appliances. Plan a house tour with your family so that they know what to do if water emergencies arise when they are alone in the house.

Start with the main water line valve and be sure it operates freely. Apply the tag and if it is especially hard to find, place a second tag in a more visible spot.

Continue the procedure every place where water is used, and see if the valves open and close easily. Over a period of time a valve can become "frozen" if not used. Usually a wrench applied to the control wheel will free up the valve; do this carefully to avoid breaking the control head. After moving and freeing the valve, check for possible leaks around the stem. Minor leakage can be stopped by applying a wrench to the cap or packing nut. Finally, apply the tags, such as "hot water kitchen," or "cold water kitchen," etc., until all valves are identified.

Closing the main water shutoff valve before leaving for a vacation is sometimes recommended, but you must be very careful about it. We prefer to recommend you have a neighbor periodically check your house. In cold weather, for example, turn-off of water might cause a freeze-up in a hot water system and a malfunction of the system. We prefer daily monitoring.

*Repairing a valve* that has seen long service is not too practical and replacement is the answer, which is especially true of gate valves and globe valves (see Figure 1–13). If you have a compression stop (Figure 1–13) that will not shut off, replace the washer, but first remove the old washer and scrape away any residue. Install the new washer and tighten the screw securely (washers are the same as in faucets). If the valve continues to leak, dress the valve seat with a reseating tool, again the same technique as used on faucets. Use the tool lightly to prevent damaging the surface and be sure to remove metal chips before reassembling.

To replace stem packing, use asbestos cord-type packing. Simply loosen packing nut (Figure 1–13), lift it up against the handle, wind the cord packing around the spindle of the stem and tighten packing nut.

## LEAKING PIPES

A leak at a threaded connection can often be stopped by unscrewing the fitting and applying pipe joint compound that will seal the joint when the fitting is screwed back together again. Small leaks in a pipe can often be repaired with a rubber patch and metal clamps or a sleeve. This must be considered as an emergency repair job and should be followed by permanent repair as soon as possible.

Leaks in tanks, such as water heaters, are usually caused by corrosion. Sometimes a safety valve may fail to open and the excessive pressure will spring a leak. While a leak may occur at only one place in a tank wall, the wall may also be corroded thin in other places. Therefore, any repair must be considered as temporary, and the tank should be replaced as soon as possible.

## GET THE NOISE OUT OF YOUR PLUMBING

There are several types of plumbing system noises, and the cause and remedy for each is treated separately here.

*Faucet noises* can be caused by a worn washer, which allows water to leak or drip (see Figure 1–10), or they can be caused by a defective internal assembly which results in chattering or whistling. Often a loose washer causes chattering by alternately stopping and then freeing the water supply when the faucet isn't shut completely off. The remedy is to disassemble the faucet and tighten the small screw holding the washer on the end of the spindle. If the washer is worn, replace it.

Chattering and whistling also occur in faucets of poor design because of the small restricted waterway. Hydraulic phenomena which

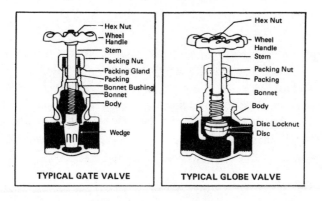

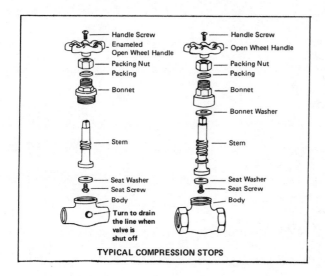

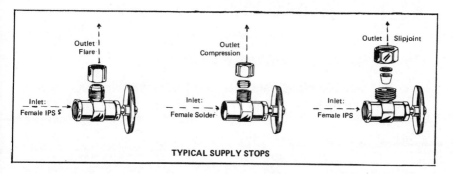

**Figure 1-13**

**Repairing leaking valves.**

occur as a result cause the noise. In that event, faucet replacement is the answer.

*Water hammer noise*—the pounding of pipes and shuddering of fixtures—is caused by a shock wave that results when the flow of water in a pipe is suddenly stopped by the rapid closing of a faucet. If allowed to continue it can cause serious damage to your plumbing, check all exposed water pipes to be sure there are no long lengths of pipe which are not fastened to a joint or beam with a pipe strap, pipe hanger or similar fastening device shown in Figure 1–14.

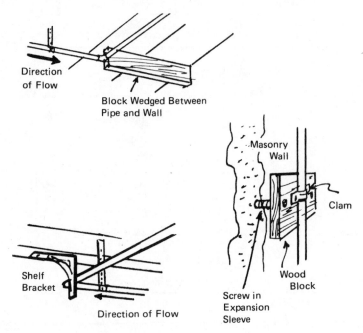

**Figure 1-14**

**Reducing pipe noise.**

Another more extreme remedy may be a professional plumber's job. It involves installing short lengths of closed pipe in the plumbing system (see Figure 1–15) which will form air chambers. These air chambers act as a cushion to absorb the excessive pressure caused in the system when a flow of water is suddenly stopped. If your system already has these air chambers, it may be that the air has leaked out. In that case draining the water system will restore the air cushion.

Toilet noises are covered under the section on leaking toilets in this chapter.

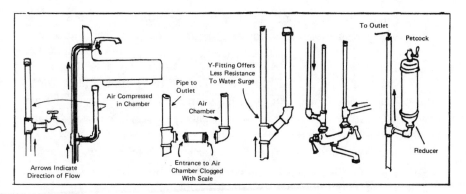

**Figure 1-15**

**Water hammer and how to deal with it.**

*Water heater tank noises* are caused when water is heated too hot, too quickly. For general household uses, water should be 130 to 140°.

Other water system noises caused by the normal functioning of your plumbing system usually are not objectionable, although sometimes they are amplified by a ceiling or a wall to a point where they become bothersome. This noise may be described as a hissing sound. Wrapping the pipes, or using sound-insulating material will help if these "normal" noises are annoying.

If you have air chambers in your line, check to see if they are free of scale. Clean and replace, if necessary, and be sure to tighten all connections to prevent air loss (see Figure 1–15).

### FROZEN PIPES

You can wrap pipe with cloth which is saturated with boiling water, or use a blowtorch if the pipe is located in a safe area, protecting walls, etc., with asbestos board. Be sure outlets are open to allow steam to escape. It is better to adopt the slower and more conservative procedure of melting ice by the use of heavy cloths. You can also use an electric iron against the pipe or direct an electric heater or electric tape to the proper location. Be sure to apply at the end of the frozen line, not the middle, since expansion of heated water may burst the pipe. As a last resort, you can remove part of the pipe at a union, insert a smaller pipe as far as it will go and then pour boiling water in—a length of rubber tubing will also work. Be sure to allow the returned water to flow into a bucket to avoid the drainage.

You can insulate pipe exposed to freezing temperatures using materials available at your plumbing supply house. Also, drain garden hose outlet, garage faucets, swimming pool connections, etc., before winter.

## SWEATING PIPES

"Sweating" pipes and plumbing fixtures in summertime are not a sign of faulty plumbing. Due to condensation of water vapor in the air, beads of moisture will form in warm weather on any pipes and fixtures containing cold water.

Normally, when not in use, the water and fixtures will warm rapidly to room temperature and the condensation will stop. When a closet tank or other fixture continues to "sweat" for hours after it has been used, it is a sign that cold water is continuing to flow through it, possibly due to an improper adjustment of the tank valve or a leak. In this case, check for leaks. Sweating pipes can be wrapped with an insulation material which prevents the condensation and formation of moisture.

## ODORS IN THE PLUMBING SYSTEM

The well-designed and correctly installed plumbing system is odorless. Odors are most likely to arise from leaks in the waste or vent piping or from traps which have lost their water seal. Unusual odors should never be ignored. Such odors are often an indication that sewer gas is present. Sewer gas, while not deadly, is noxious and capable of causing headaches and other minor illnesses. Sewer gas is vitiated air and should be prevented from entering the house.

If it is suspected that sewer gas in entering through a leak in the piping, a plumber should be called in to test the system by means of smoke, water or peppermint to indicate the location of the leak. To clarify the function of traps and vents once again, every plumbing fixture is the terminus of the city water supply system and the beginning of a sewerage system. The faucets control the water supply. The traps and vents control the sewer air. They do so by a very simple method. Sewer air will not penetrate a water barrier. Therefore, a device is employed which keeps several inches of water between the house air and the sewer air. This is the trap which is plainly visible under such plumbing fixtures as sinks. A trap can lose its seal through evaporation, which happens only if a fixture is used very infrequently.

Also a trap would lose its water seal by siphonic action unless the air on the sewer side is balanced with the air on the house side. This is the function of the vents. Occasionally, due to changes in atmospheric conditions, a correctly vented trap will lose its seal.

## DRAINING PLUMBING IN A VACANT HOUSE

If your house must be vacated during cold weather and the heating system turned off, with this procedure you will be secure:

Shut off the water supply at the main shut-off valve at the street, then, beginning with those on the top floor, open all faucets, and leave them open. When water stops running from these faucets, open the cap in the main pipe valve in the basement, and drain the remaining water into a pail or tub. Remember that this cap must be closed until the faucets have run dry, or the house water supply will flow from this valve and flood the basement. A professional plumber can also do this for a small fee if you are in a hurry.

Remove all water in the traps under sinks, water closets, bathtubs, and lavatories by opening the clean out plugs at the bottom of traps and draining them into a pail. If no traps are provided, use a force pump or other method to siphon the water out. Sponge all the water out of the water closet bowl. Clean out all water in the flush tank.

Fill all traps with kerosene, crude glycerine, or one of the non-freeze compounds used in automobiles. This will seal off bad odors from waste pipes. Alcohol and kerosene are a recommended mixture, as the kerosene will rise to the top and prevent evaporation of the alcohol. Two quarts of this mixture will be needed for each toilet trap, but other plumbing fixtures will need less. Do not forget to fill the trap in the floor drain in the basement.

Drain all hot water tanks. Most water tanks are equipped with a vented tube at the top which lets air in and allows the water to drain out the faucet at the bottom. Make sure all horizontal pipes drain properly. Air pressure will get rid of trapped water in these pipes, but occasionally the piping may have to be disconnected and drained.

If your house is heated by hot water steam, drain the heating pipes and boiler before leaving. Fires should be completely out and the main water supply turned off at the basement wall or street. Draw off the water from the boiler by opening the draw-off valve at the lowest point in the system.

Open the water supply valve to the boiler so no water will be trapped above it. If you have a hot water system, begin with the highest radiators and open the air valve on each as fast as the water lowers. Every radiator valve must be opened on a 1-pipe system to release condensation.

*Note*: when you return home, refill your heating system *before* lighting the boiler.

# HOW TO REPLACE
# OLD FITTINGS

There comes a time when a faucet or valve simply becomes too old to mess with. It could be for esthetic reasons. Perhaps the lavatory or kitchen faucets have become pitted and/or past the point of repair. It is time, you have decided, for replacement.

The smartest thing to do when buying a replacement is to take the old one with you. Or, if that is not practical, you should have the center to center measurements exactly on single-unit fixtures. Faucet makers include installation instructions (see Figures 2–1, 2, and 3 for typical instruction directions for a twin handle lavatory faucet, a twin handle kitchen faucet, and the installation instructions for a pop-up drain assembly for Rockwell products), and most are easy enough to follow. However, you may run across fuzzy directions or non-existent ones. The illustrations you see within this chapter are fairly standard in procedure, regardless of the type of fixture you buy, and will give you a fine indication of the kind of work you will be facing. It's not complicated, really— it just takes the right tools and some patience.

A professional tip: if you are going to replace a faucet that was originally connected to iron pipe, be sure the new one is hooked up to copper tubing—it's easier to work with since you don't have to be concerned with threading the pipe.

### SHOWER AND BATH FAUCETS

There is no need to deface the bathroom wall when replacing these faucets, since you should have a rear access panel that will allow you to connect the fittings. Figure 2–4 shows a variety of fittings, including four

# installation instructions
# twin handle lavatory faucet

**NOTE: Always shut off hot and cold water supplies.**

It is recommended that plumber's pipe compound be used on all metal threads to insure proper seal.

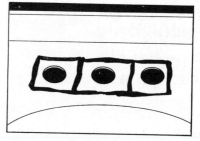

**1** Place beads of plumbers' putty on sink top.

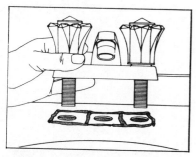

**2** Insert shanks through holes in sink.

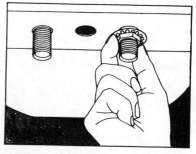

**3** Fasten faucet into place with washers and nuts. Tighten firmly.

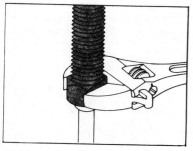

**4** Connect shanks to water supply lines.

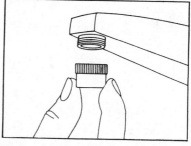

**5** Remove aerator assembly located at the end of the spout and flush the faucet of foreign matter . . . then replace aerator.

**your lavatory faucet is now completely installed and ready for years of guaranteed satisfaction.**

**Figure 2-1**

# installation instructions

## twin handle kitchen faucet

**NOTE: Always shut off hot and cold water supplies.**

It is recommended that plumber's pipe compound be used on all metal threads to insure proper seal.

**For models with spray, start with step 1.**

**For models less spray, start with step 3.**

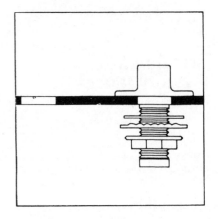

**1** Place spray holder through faucet hole at extreme right of sink. Add fiber washer, corrugated washer and locknut. Tighten firmly.

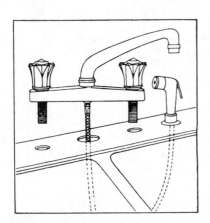

**2** Place spray hose through spray holder and up through center hole of sink ledge. Attach spray hose to body.

**3** Place beads of plumbers' putty on sink top.

**Figure 2-2**

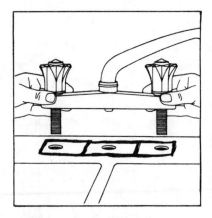

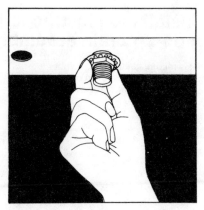

**4** Place putty cup over shanks, insert faucet into holes and press firmly into position.

**5** Install lock washers and locknuts to shanks and tighten firmly.

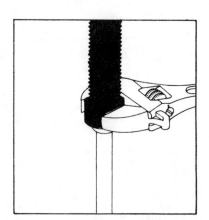

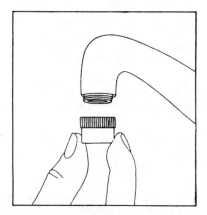

**6** Connect shanks to water supply lines.

**7** Remove aerator assembly located at the end of the spout and flush the faucet of foreign matter . . . then replace aerator.

**your kitchen faucet is now completely installed**

**Figure 2-2 (cont.)**

# installation instructions
# pop-up drain assembly

It is recommended that plumber's
pipe compound be used on all metal
threads to insure proper seal.

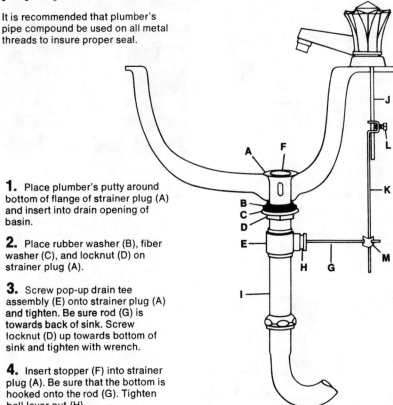

**1.** Place plumber's putty around
bottom of flange of strainer plug (A)
and insert into drain opening of
basin.

**2.** Place rubber washer (B), fiber
washer (C), and locknut (D) on
strainer plug (A).

**3.** Screw pop-up drain tee
assembly (E) onto strainer plug (A)
and tighten. Be sure rod (G) is
towards back of sink. Screw
locknut (D) up towards bottom of
sink and tighten with wrench.

**4.** Insert stopper (F) into strainer
plug (A). Be sure that the bottom is
hooked onto the rod (G). Tighten
ball lever nut (H).

**5.** Insert unthreaded end of
tailpiece (I) into trap and screw
other end up into main body of
strainer plug (A). Tighten trap nut.

**6.** Insert upper lift rod (J) through
hole in rear of faucet into the lower
lift rod (K) and tighten set screw (L).

**7.** Place lever rod (G) through
proper opening in the lower lift rod
and adjust spring clip (M) to hold
lower lift rod in place. Installation
is complete.

**A** Strainer Plug
**B** Rubber Washer
**C** Fiber Washer
**D** Locknut
**E** Drain Tee Assembly
**F** Stopper
**G** Ball Lever Rod
**H** Ball Lever Nut
**I** Tailpiece
**J** Upper Lift Rod
**K** Lower Lift Rod
**L** Set Screw
**M** Spring Clip

**NOTE:** You can adjust assembly for maximum opening by raising or
lowering lift rod by resetting screw.

**Figure 2-3**

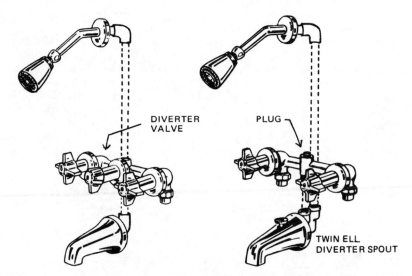

This is a 3-Valve Diverter with Shower Head and Spout.

This is a 2-Valve Diverter with Shower Head and Twin Ell Diverter Spout.

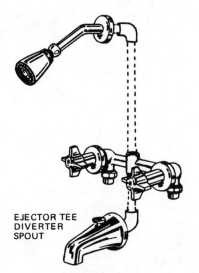

This is a 2-Valve Diverter with Shower Head and Ejector Tee Diverter Spout.

**Figure 2-4**

types of shower diverters, heads and spouts, and two types of tub faucets. Figure 2–5 shows how supply lines are hooked up with either threaded or solder connections. For tips on making these connections see Chapter 6—Working With Pipe and Tubing.

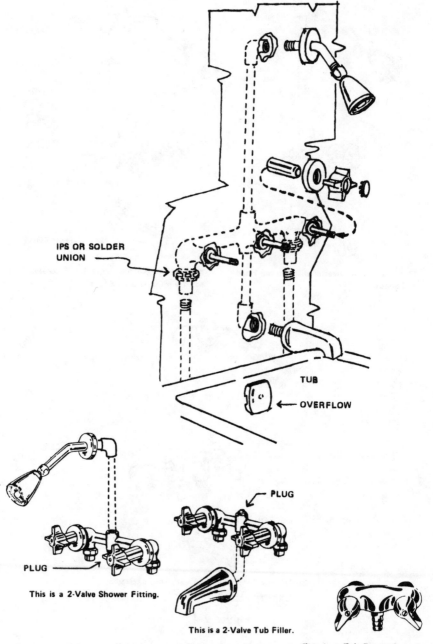

IPS OR SOLDER
UNION

TUB

OVERFLOW

PLUG

PLUG

This is a 2-Valve Shower Fitting.

This is a 2-Valve Tub Filler.

This is a Tub Faucet for a Leg-type Tub.

**Figure 2-5**

Both two and three-valve bath and shower fittings are made with 8″ centers, which is the most common measure.

*Bath drains* are of two types—trip lever and pop-up—in the modern area of plumbing. There is also a connected waste and overflow system with the good old chain and rubber stopper most of us remember from the old days (see Figure 2–6 for these types). They do what they say. The pop up type has a drain plug which actually lifts when the lever is shifted; the

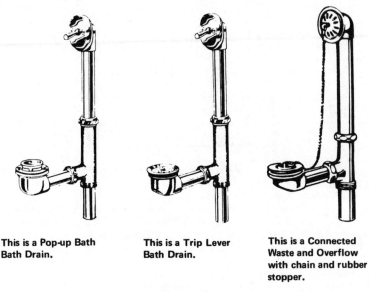

This is a Pop-up Bath Bath Drain.

This is a Trip Lever Bath Drain.

This is a Connected Waste and Overflow with chain and rubber stopper.

**Figure 2-6**

trip lever type has an internal mechanism, and the good old chain and stopper are hard to improve on mechanically. The trip lever mechanism can be seen in Figure 2–7, while the mechanism of the chain and stopper assembly is seen in Figure 2–8.

Both the trip lever and pop-up bath drains come in 1½″ diameter and are usually made of strong 17-guage tubing. When you buy the assembly, the cast brass nuts, strainer plug and tailpiece are supplied, and you will find that the lift rod is adjustable to fit all modern tubs. The connected waste and overflow bath drains are made in 1½″ and 1⅜″ diameter and the chain and stopper and tailpiece are supplied.

Figure 2–9 shows the lower assembly of a tub drain, with pipe lengths varying to suit the measure of your particular tub.

Other views of replacing old plumbing fittings in the lavatory are seen in Figure 2–10, as is also a cross section of a typical installation, and

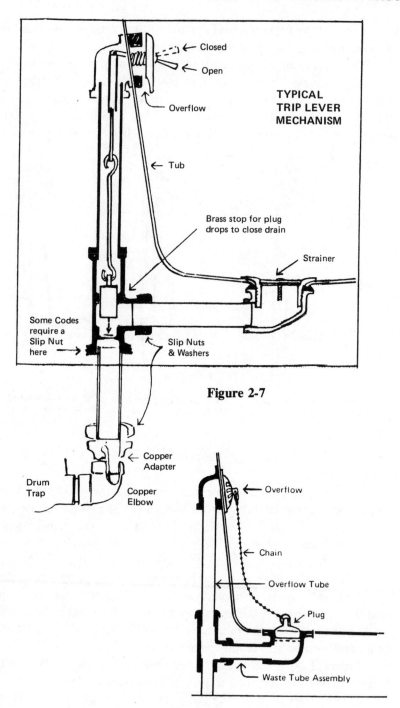

**Figure 2-7**

**Figure 2-8**

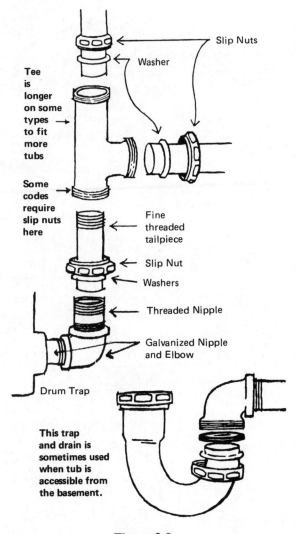

**Figure 2-9**

one showing a typical lavatory faucet with pop-up drain in operation—these are courtesy of Rockwell, which makes many such fittings.

Drain traps for sinks and lavatories may be a nuisance to replace, but they are really not overly complicated, as you will note in Figure 2–12, which shows a typical "P" trap installation, typical wall connections for traps, a typical "S" trap installation, and instructions on replacing a typical trap bend.

*Here's how to replace a typical ballcock in your toilet tank*, as well

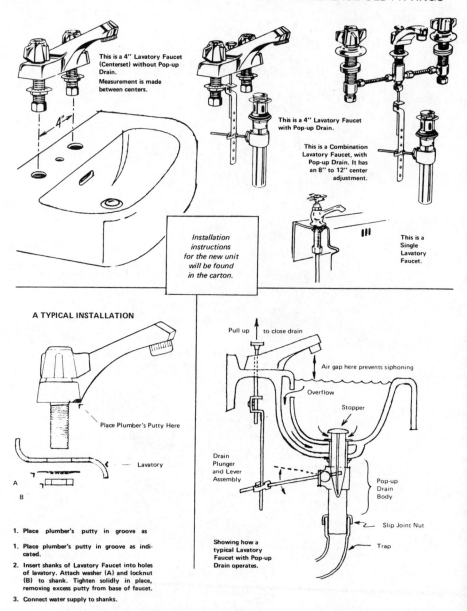

This is a 4" Lavatory Faucet (Centerset) without Pop-up Drain.
Measurement is made between centers.

This is a 4" Lavatory Faucet with Pop-up Drain.

This is a Combination Lavatory Faucet, with Pop-up Drain. It has an 8" to 12" center adjustment.

Installation instructions for the new unit will be found in the carton.

This is a Single Lavatory Faucet.

A TYPICAL INSTALLATION

Place Plumber's Putty Here

Lavatory

A

B

1. Place plumber's putty in groove as

1. Place plumber's putty in groove as indicated.

2. Insert shanks of Lavatory Faucet into holes of lavatory. Attach washer (A) and locknut (B) to shank. Tighten solidly in place, removing excess putty from base of faucet.

3. Connect water supply to shanks.

Pull up  to close drain

Air gap here prevents siphoning

Overflow

Stopper

Drain Plunger and Lever Assembly

Pop-up Drain Body

Slip Joint Nut

Trap

Showing how a typical Lavatory Faucet with Pop-up Drain operates.

**Figure 2-10**

as a typical flush valve. First (see Figure 2–11) you remove the coupling nut and the locknut from the shank. Then you place the shank of the ballcock through the opening in the tank with the large rubber cone washer in place, as seen in the illustration. After connecting water supply to the shank, you slip one end of the plastic refill tube over the serrated

**Replacing a Typical Ballcock**

**Replacing a Typical Flush Valve**

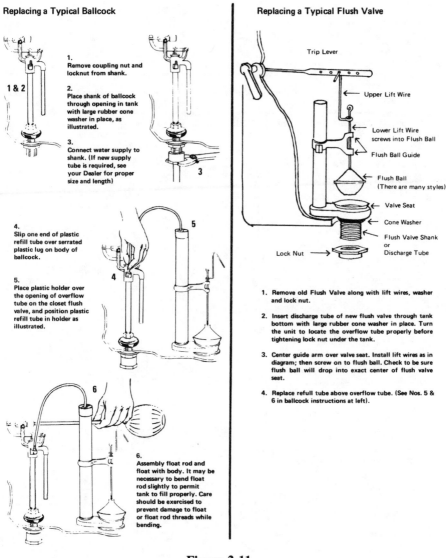

**1.**
Remove coupling nut and locknut from shank.

**2.**
Place shank of ballcock through opening in tank with large rubber cone washer in place, as illustrated.

**3.**
Connect water supply to shank. (If new supply tube is required, see your Dealer for proper size and length)

**4.**
Slip one end of plastic refill tube over serrated plastic lug on body of ballcock.

**5.**
Place plastic holder over the opening of overflow tube on the closet flush valve, and position plastic refill tube in holder as illustrated.

**6.**
Assembly float rod and float with body. It may be necessary to bend float rod slightly to permit tank to fill properly. Care should be exercised to prevent damage to float or float rod threads while bending.

Trip Lever

Upper Lift Wire

Lower Lift Wire screws into Flush Ball

Flush Ball Guide

Flush Ball (There are many styles)

Valve Seat

Cone Washer

Flush Valve Shank or Discharge Tube

Lock Nut

1. Remove old Flush Valve along with lift wires, washer and lock nut.

2. Insert discharge tube of new flush valve through tank bottom with large rubber cone washer in place. Turn the unit to locate the overflow tube properly before tightening lock nut under the tank.

3. Center guide arm over valve seat. Install lift wires as in diagram; then screw on to flush ball. Check to be sure flush ball will drop into exact center of flush valve seat.

4. Replace refull tube above overflow tube. (See Nos. 5 & 6 in ballcock instructions at left).

**Figure 2-11**

plastic lug on the body of the ballcock, place the plastic holder (as seen in Figure 2–11) over the opening of the overflow tube on the closet flush valve, and position the plastic refill tube in the holder as shown. The float rod is then assembled to float properly, and it may be necessary to bend the float rod a bit to allow the tank to fill properly, with care taken to

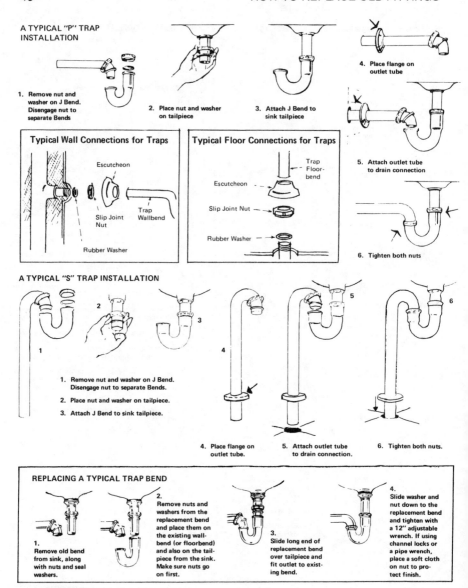

**Figure 2-12**

prevent damage to the float or rod threads while bending, as in Figure 2–11.

   *To replace a typical flush valve*, examine Figure 2–13 carefully. An important point is to check to be sure that the flush ball will drop into the exact center of the flush valve seat, and to be sure to replace the refill tube over the overflow tube.

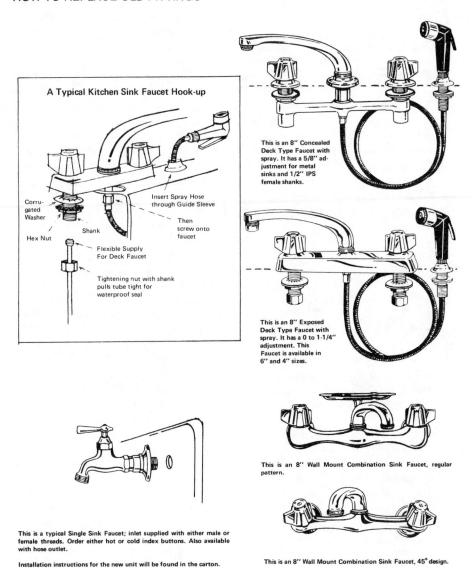

A Typical Kitchen Sink Faucet Hook-up

Corrugated Washer

Hex Nut

Shank

Insert Spray Hose through Guide Sleeve

Then screw onto faucet

Flexible Supply For Deck Faucet

Tightening nut with shank pulls tube tight for waterproof seal

This is an 8" Concealed Deck Type Faucet with spray. It has a 5/8" adjustment for metal sinks and 1/2" IPS female shanks.

This is an 8" Exposed Deck Type Faucet with spray. It has a 0 to 1-1/4" adjustment. This Faucet is available in 6" and 4" sizes.

This is an 8" Wall Mount Combination Sink Faucet, regular pattern.

This is a typical Single Sink Faucet; inlet supplied with either male or female threads. Order either hot or cold index buttons. Also available with hose outlet.

Installation instructions for the new unit will be found in the carton.

This is an 8" Wall Mount Combination Sink Faucet, 45° design.

**Figure 2-13**

*Kitchen sink hookups* are not that much different than lavatory ones, with the exception that you will have a traversing faucet with a spray attachment. See Figure 2–13, which also shows a typical single sink faucet.

*Replacing a kitchen sink strainer* may require one or two men, depending on the type (see Figure 2–14). Important points include clean-

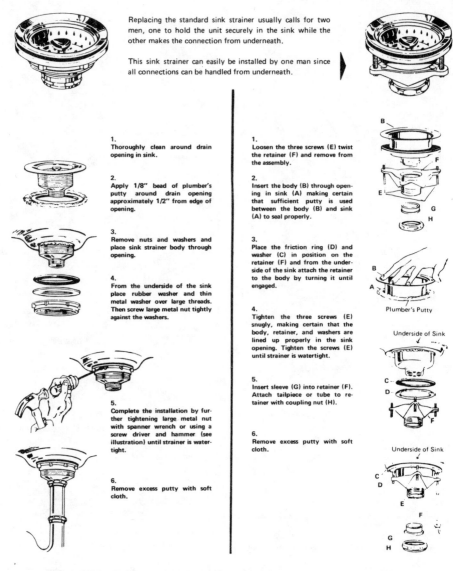

Replacing the standard sink strainer usually calls for two men, one to hold the unit securely in the sink while the other makes the connection from underneath.

This sink strainer can easily be installed by one man since all connections can be handled from underneath.

**1.**
Thoroughly clean around drain opening in sink.

**2.**
Apply 1/8" bead of plumber's putty around drain opening approximately 1/2" from edge of opening.

**3.**
Remove nuts and washers and place sink strainer body through opening.

**4.**
From the underside of the sink place rubber washer and thin metal washer over large threads. Then screw large metal nut tightly against the washers.

**5.**
Complete the installation by further tightening large metal nut with spanner wrench or using a screw driver and hammer (see illustration) until strainer is watertight.

**6.**
Remove excess putty with soft cloth.

**1.**
Loosen the three screws (E) twist the retainer (F) and remove from the assembly.

**2.**
Insert the body (B) through opening in sink (A) making certain that sufficient putty is used between the body (B) and sink (A) to seal properly.

**3.**
Place the friction ring (D) and washer (C) in position on the retainer (F) and from the underside of the sink attach the retainer to the body by turning it until engaged.

**4.**
Tighten the three screws (E) snugly, making certain that the body, retainer, and washers are lined up properly in the sink opening. Tighten the screws (E) until strainer is watertight.

**5.**
Insert sleeve (G) into retainer (F). Attach tailpiece or tube to retainer with coupling nut (H).

**6.**
Remove excess putty with soft cloth.

Plumber's Putty

Underside of Sink

Underside of Sink

**Figure 2-14**

ing around the drain opening, removing the nuts and washers and placing the sink strainer body through the opening for a proper fit, and making certain that the body, retainer and washers of the strainer are lined up properly in the sink opening. Be sure to tighten the screws until the strainer is watertight, but do not overtighten.

Continuous wastes can be had two ways; either center or end outlets

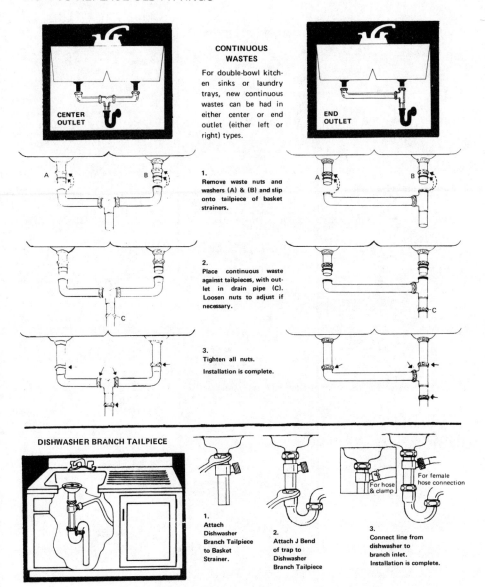

**CONTINUOUS WASTES**

For double-bowl kitchen sinks or laundry trays, new continuous wastes can be had in either center or end outlet (either left or right) types.

1.
Remove waste nuts and washers (A) & (B) and slip onto tailpiece of basket strainers.

2.
Place continuous waste against tailpieces, with outlet in drain pipe (C). Loosen nuts to adjust if necessary.

3.
Tighten all nuts.

Installation is complete.

CENTER OUTLET

END OUTLET

DISHWASHER BRANCH TAILPIECE

1.
Attach Dishwasher Branch Tailpiece to Basket Strainer.

2.
Attach J Bend of trap to Dishwasher Branch Tailpiece

3.
Connect line from dishwasher to branch inlet. Installation is complete.

For hose & clamp

For female hose connection

NOTE: A dishwasher branch tailpiece can often be connected to a garbage disposal.

**Figure 2-15**

(left or right) are available, as seen in Figure 2–15. Dishwasher branch tailpieces can be installed without much of a problem to the kitchen sink drain, as also seen in Figure 2–15, and can also be made to serve a garbage disposal unit.

*There are three ways to replace supply stops, depending upon the*

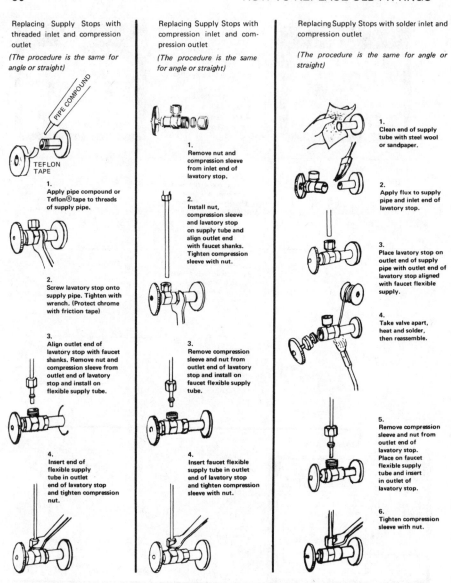

Replacing Supply Stops with threaded inlet and compression outlet

*(The procedure is the same for angle or straight)*

**1.**
Apply pipe compound or Teflon® tape to threads of supply pipe.

**2.**
Screw lavatory stop onto supply pipe. Tighten with wrench. (Protect chrome with friction tape)

**3.**
Align outlet end of lavatory stop with faucet shanks. Remove nut and compression sleeve from outlet end of lavatory stop and install on flexible supply tube.

**4.**
Insert end of flexible supply tube in outlet end of lavatory stop and tighten compression nut.

Replacing Supply Stops with compression inlet and compression outlet

*(The procedure is the same for angle or straight)*

**1.**
Remove nut and compression sleeve from inlet end of lavatory stop.

**2.**
Install nut, compression sleeve and lavatory stop on supply tube and align outlet end with faucet shanks. Tighten compression sleeve with nut.

**3.**
Remove compression sleeve and nut from outlet end of lavatory stop and install on faucet flexible supply tube.

**4.**
Insert faucet flexible supply tube in outlet end of lavatory stop and tighten compression sleeve with nut.

Replacing Supply Stops with solder inlet and compression outlet

*(The procedure is the same for angle or straight)*

**1.**
Clean end of supply tube with steel wool or sandpaper.

**2.**
Apply flux to supply pipe and inlet end of lavatory stop.

**3.**
Place lavatory stop on outlet end of supply pipe with outlet end of lavatory stop aligned with faucet flexible supply.

**4.**
Take valve apart, heat and solder, then reassemble.

**5.**
Remove compression sleeve and nut from outlet end of lavatory stop. Place on faucet flexible supply tube and insert in outlet of lavatory stop.

**6.**
Tighten compression sleeve with nut.

**Figure 2-16**

*construction of the stop*: Figure 2–16 illustrates how to replace a supply stop with a threaded inlet and compression outlet, how to replace one with a compression inlet and compression outlet, and one with solder inlet and a compression outlet. You will note that the procedure is the same for an angled or a straight stop.

Replacing Hose Faucets, Lawn Faucets, Utility Faucets, and Boiler Drains with male or female IP threads

**1.**
Remove dirt and chips from pipe ends. Threads should be clean and smooth.

Replacing Faucets with solder ends

**1.**
Clean ends of copper pipe thoroughly.

**2.**
Remove stem to protect seat washer and apply heat to pipe.

**2.**
Apply pipe compound or Teflon tape to pipe threads.

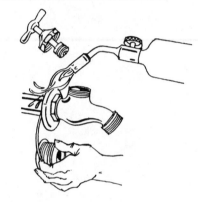

**3.**
Tighten faucet as much as possible by hand.

**3.**
Apply solder. (See Chapter 6, WORKING WITH PIPE.)

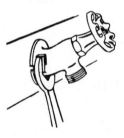

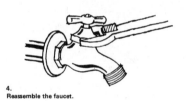

**4.**
Reassemble the faucet.

**4.**
Complete installation by tightening faucet onto pipe with wrenches. (2 or 3 threads of pipe should be left exposed when valve is tight, provided pipe was properly threaded).

**5.**
Remove excess pipe compound with cloth.

**Figure 2-17**

*Replacing hose and utility faucets is relatively easy*, as seen in Figure 2–17. It's really a matter of careful unscrewing and rescrewing, with cleanliness of the pipe ends kept in mind, and appropriate applications of pipe compound or tape. For working with faucets which require solder, see in addition to Figure 2–17 the chapter titled Working With Pipe and Tubing (Chapter 6).

*Valve replacement* is demonstrated in Figure 2–18. Again, judicious

Replacing Gate Valves, Globe Valves, Compression Stops and other Valves with threaded connections.

**1.**
Remove dirt and chips from pipe ends. Threads should be clean and smooth.

**2.**
Apply pipe compound or Teflon Tape to pipe threads.

**3.**
Tighten valve as much as possible by hand. (Arrow on valve indicates direction of flow).

**4.**
Complete installation by tightening valve onto pipe with wrenches. (2 or 3 threads of pipe should be left exposed when valve is tight, provided pipe was properly threaded).

**5.**
Remove excess pipe compound with cloth.

**6.**
If valve has a drain, be sure it is closed before turning on water supply.

Replacing Gate Valves, Globe Valves, Compression Stops and other Valves with solder connections.

**1.**
If new installation, remove all burrs and rough edges from copper tubing.

**2.**
Brighten copper tubing and valve opening with emery cloth or steel wool.

**3.**
Apply non-corrosive flux to copper tubing, slip valve over tubing and rotate several turns to spread flux evenly. (Arrow on valve must point in direction of water flow).

**4.**
Remove valve stem (excessive heat may damage seat washer) and apply heat to all sides of valve with propane torch.

**5.**
When flux begins to bubble, apply 50-50 wire solder to edge of valve opening until thin line of solder forms around face of opening.

**6.**
Remove excess solder with a clean rag or brush. Check joint for leakage.

**7.**
If valve has a drain, be sure it is closed before turning on water supply.

**Figure 2-18**

use of pipe compound and Teflon tape, and careful tightening are the secrets. Do note that if the valve to be replaced has a drain, be sure it is closed before turning on the water supply. If you are working with a solder connection rather than a threaded connection, you will want to remove burrs and rough edges with emery cloth or steel wool from the copper tubing. Here you will need a propane torch and some solder, and

This is a typical Combination Laundry Tray Faucet with 3¼" centers, integral clamps, and 6" spout.

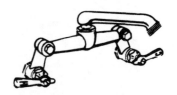

This is a "California" Combination Laundry Tray Faucet with 4" to 9" adjustable centers.

This is a typical Combination Laundry Tray Faucet with 4" centers and 2-way brackets for wall or tub mount installation.

This is a typical built-in Washing Machine Filler Valve with 8" centers.

This is a typical Combination Laundry Tray Faucet with 4" centers and post brackets.

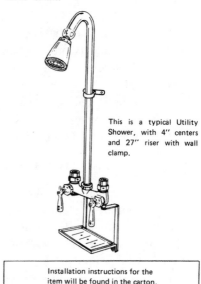

This is a typical Utility Shower, with 4" centers and 27" riser with wall clamp.

This is a typical 4" Ledge Type Laundry Tray Faucet with 0 to 1¼" adjustment.

Installation instructions for the item will be found in the carton.

**Figure 2-19**

do a careful soldering job. Patience, patience and some practice and you, too, can save $25 an hour by doing these relatively simple chores!

A variety of laundry room faucets are shown in Figure 2–19, to show you what is available at your plumbing supply center. These models are from Rockwell, and each has installation instructions in the carton.

# UNDERSTAND YOUR PLUMBING SYSTEM— HOW IT WORKS

There are three basic parts to your plumbing system: the water supply system, the fixtures in which water is put to use, and the drainage system. Wherever you live, these three parts are necessary to your full enjoyment of your home. Home plumbing is not a mystery, nor is it overly complicated—it is simply there to make it convenient for you to use water the ways you desire. And—lucky for you—it is one of the easier trades for a homeowner to learn.

## THE WATER SUPPLY SYSTEM

An adequate capacity (volume and correctly placed outlets) is the most important system feature; long, trouble-free life is second. If you buy water from a utility your system begins at the utility's main (distribution pipe that serves many users). See Figure 3-1 for a diagram of a typical water supply system. If you have a private source of water your supply system actually includes the source (well, lake, etc.) and the pump, mains and other parts needed to deliver water into your home. In either case, however, for our immediate purpose we shall consider your water supply system to consist only of the piping directly associated with your home.

Hence, your system begins with a building main (BM)—which may contain a utility meter—of ample size or diameter to carry water at the maximum rate (gallons per minute) at which you will need it. Inside the building this line branches to become two lines. One is the cold water

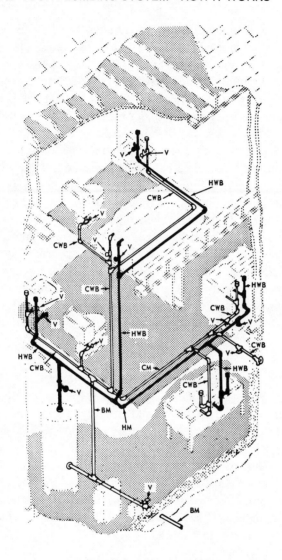

**A TYPICAL WATER-SUPPLY SYSTEM**

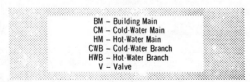

BM – Building Main
CM – Cold-Water Main
HM – Hot-Water Main
CWB – Cold-Water Branch
HWB – Hot-Water Branch
V – Valve

**Figure 3-1**

main (CM), which winds around through the building as necessary to supply all of your cold-water outlets. The other goes directly to your hot-water heater, from which it emerges as the hot-water main (HM). Thus your house piping really consists of two separate systems—a cold and a hot—which run in parallel to all fixtures which use both.

Since fresh water is supplied under pressure (is moved by a pump), it will flow upward as well as down. The cold and hot water pipes therefore can follow any path of convenience through the building walls or floors, etc., and the most direct and economical routes are generally used. When practical, each main will have as many separate branches (CWB and HWB in Figure 3–1) as there are fixtures to be served, and each branch is a smaller pipe size than the main (which usually is the same size up to the point where the last two branches take off). However, a branch may serve two or more fixtures, in which case it is sized accordingly. Portions of mains or branches that go up or down are referred to as risers; two or more lengths of pipe are joined together by fittings. If desired, your system may include a water softener or some type of filter—or the cold-water main may be extended to serve outdoor faucets or a lawn-sprinkler system.

## FIXTURES

A fixture is any appliance (tub, shower, toilet, sink, dishwasher, etc.) that uses (or helps you use) water. Those piping parts (faucets, traps, etc.) which are exposed to view—and, usually, are chrome plated—are considered part of the fixture (though they may not be furnished with the fixture), rather than parts of the water supply or drainage systems. Remember these trim parts when ordering fixtures. Fixtures are the most obviously important part of any plumbing installation. The selection you make will determine—for years to come—how much satisfaction you will enjoy. Most of them will become a built-in and costly to change part of the building, so choose them wisely.

## THE DRAINAGE SYSTEM

Sufficient capacity and pitch, tight sealing, proper venting, and provisions for clean-out all are important to a good drainage system. See Figure 3–2 for a diagram of a typical drainage system, with an explanatory code.

Used water—and the waste it carries with it—must be disposed of, both for convenience and for the sake of your health. Waste creates gases that are unpleasant and often harmful which also must be dispelled. Your drainage system therefore has two functions: to dispose of water and solid

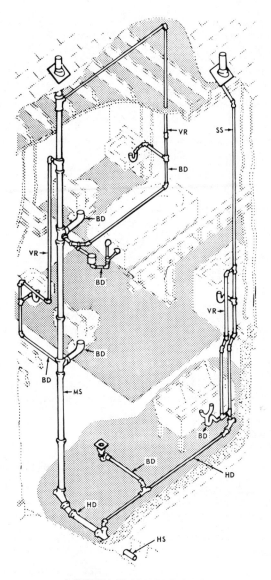

A TYPICAL DRAINAGE SYSTEM

MS — Main (Soil) Stack
SS — Secondary (Soil) Stack
BD — Branch Drain
HD — House Drain
HS — House Sewer
VR — Vent Run

**Figure 3-2**

wastes to a municipal sewer or septic tank, cistern or disposal field, and to dispel the noxious gases into air you won't be breathing.

There are no pumps to force water away from the fixtures. Waste water flows by gravity alone. All drainage parts must be pitched or sloped downhill—and must be large enough in diameter and smooth enough inside to prevent accumulation of solids at any point. These requirements make it uneconomical to have any unnecessary drainage parts. Therefore, a whole plumbing installation generally is planned around the drainage system, to drain as many as possible of the fixtures into the same main pipe (and conserve on main pipes). This also makes the venting or dispelling to the air of gases more economical.

A drain pipe (vertical or nearly so) that collects waste from one or more fixtures is called a soil stack (or stack). If a toilet drains into it, it is a main stack (MS in Figure 3–2). Every home has one main stack and may have several if different ones are required to drain several toilets. If no toilet drains into a stack it can be smaller (pipe diameter) than a main stack—and is called a secondary stack (SS). Each fixture is joined to its stack by a branch drain (BD) that must slope at a down angle to the stack. All stacks extend down into a crawlspace or basement, or into the ground under the house or basement, and connect with a horizontal (sloped) run called the house drain (HD). Usually, there is one house drain, but there can be several. In either case, the house drain(s) becomes a house sewer (HS) on the outside of the building—and continues on horizontally to connect with the final disposal (term used to define a public sewer or private septic tank).

Gases float upwards—and will float up through the same pipes that carry water down. Water drainage pipes are used insofar as possible to vent pipes. To this end, each soil stack usually is extended upward through the roof where its open top will allow gases to rise and be dissipated in the air above. Some codes permit this type of venting alone; others require that certain fixtures (depending upon type or location) be revented through the roof or back to the soil stack by additional pipe that is pitched upward (not downward like the branch drain) from the fixture. System branches used for this purpose are called vent runs (VR).

Drainage and vent pipes are sized (diameter of pipe) according to the burden each must carry. There is a limit to the distance between any toilet and its soil stack—and each is connected to its stack by a special branch drain called a closet bend. All other fixtures must have some kind of water trap installed just below the fixture outlet—to block the branch drain (by remaining always filled with water) so that gases cannot rise out of the outlet. Standard fittings are used in vent runs; but all fittings that carry water must be the type (called sanitary) designed for drainage. There should be a cleanout ferrule with a removable plug at the foot of each stack to permit use of a cable, if necessary, to unstop the connected house drain and the sewer.

## THE CODES

With health at stake, all plumbing conforms to safety standards which apply to both new and repair work. In larger cities, detailed codes or standards have become part of the law and permits are required before work is started. After completion the work must be inspected and approved. Some cities require that plumbing can only be handled by licensed plumbers; others permit any individual to do the work, but it must pass official inspection. It is always advisable to check the codes that are in effect in your area.

## SUMMARIZING AN INSTALLATION

Enjoyment of your plumbing is your first consideration. If you want or must have a bathroom, sink or automatic washer in a certain location, this is where it should be—if economics and practical considerations make it possible. So start your planning with your wishes—but also try to allow for a practical solution.

As said before, the planning of a plumbing system usually centers around the drainage requirements. For economy it is preferable to have as few soil stacks as possible (many homes have only one), and to locate all fixtures so they can drain into the stack or stacks decided upon. Once the stack location(s) and fixture arrangements have been determined, the water supply is no great problem. There are many ways of running water lines to fixtures.

# 4

# WHAT YOU SHOULD KNOW ABOUT PLUMBING TOOLS

In addition to the tools suggested for your Plumbing First Aid Kit and to those you have around the house, there are a few "specials" you should have if you are to make neat and quick repairs and installations.

You can either purchase these "specials" or rent as many of them as the need arises.

Before investing in tools, note if your system has copper piping calling for solder sealed connections of whether iron pipe has been used. Each type calls for a different set of tools.

Also, for certain jobs, you will need a blowtorch and a vessel to melt lead needed for joint sealing.

## TOOLS FOR IRON PIPE WORK

### Stillson wrenches (pipe wrenches)

Generally, two are required, one for holding, the other for turning (see Figure 4–1). To work with pipe in the ⅛" to ¾" range, you will want wrenches for 6", 8", 10" length; for pipe in the ½" to 1½" range, the wrench should be 14"; for pipe up to 2", the wrench should be 18". For larger pipe sizes (big drain pipes) chain wrenches are available. To use, simply wrap the chain around the pipe and draw it tight by pulling the handle. You should not use wrenches that "bite" into the metal when tightening or removing nuts. Nor should these wrenches be used on any type fitting whose surface can be easily marred, including tubing or pipe with chrome finish.

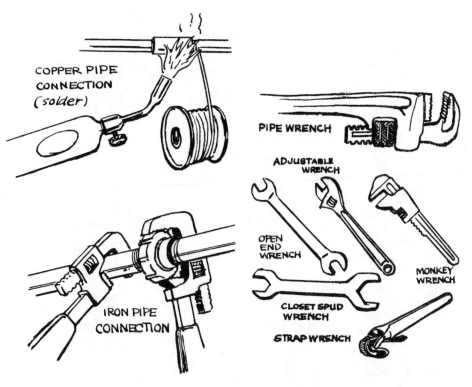

**Figure 4-1**

*Monkey wrenches, open-end wrenches, and adjustable (crescent) wrenches* are used for square or hex nuts and for working with the interior parts of faucets and valves.

*Closet-spud wrenches* are "specials" in that they are thin and can easily fit in tight places.

*Vise-grip wrenches* are extremely handy when working with small diameter pipe sizes (see Figure 4-2).

*Basin wrenches* are used to remove or tighten supply nuts and hose coupling nuts on faucet spray attachments under sinks and lavatories.

*Socket wrenches for packing nuts* are available in many sizes to fit tub and shower valves. Their shape is hex and they fit over the faucet stem to remove the valve packing nuts and stem assemblies. Since these nuts and assemblies are made of brass and can be broken easily, they should not be removed with ordinary open-end or adjustable wrenches.

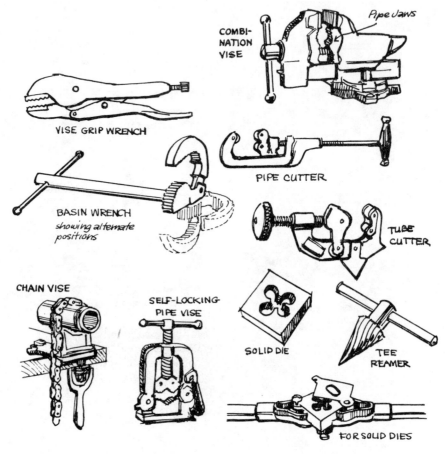

**Figure 4-2**

## TOOLS FOR THREADING PIPE

**Dies**

Generally, you can buy pipe threaded to order. To thread your own, you will need dies for cutting the external threads. Dies are made in the identical size as the fittings to which the threaded pipe will connect and are securely held in a die stock which has a provision for long handles required to turn the tool, as in Figure 4–3. In cutting threads, it is important to hold the pipe securely in a vise and to use the proper type of cutting oil.

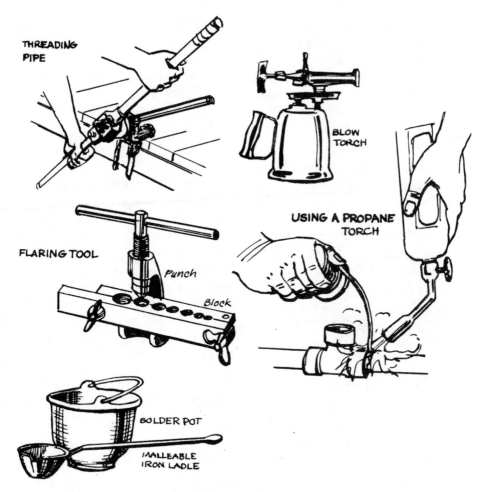

**Figure 4-3**

## Plumber's vises

These differ from machinists' vises in that they have serrated jaws to prevent pipe from turning.

There are many types of vises available. Just be sure the one you select will hold the pipe securely and is large enough to handle all the sizes you will be working with.

Generally, plumbers use a bench-type vise with v-shaped jaws that grip the pipe from both top and bottom. The upper jaw is adjustable, the lower jaw is fixed.

Another type of vise found very satisfactory is the chain vise. It is easier and quicker to use and holds pipe securely. Its operation is similar to a chain wrench.

Combination vises are available and can be used in emergencies, but as a rule, are not recommended for heavy duty work.

## CUTTING PIPE

A pipe cutter will cut pipe quicker and more accurately than a hacksaw. To use, simply start the cutter over the pipe and, as you revolve it, tighten the handle slightly. Always use thread cutting oil on both cutter wheel and pipe.

After pipe has been cut, it will be necessary to use a reamer to clear away burrs and rough surfaces. Reamers are available in two types, fluted and spiral. The fluted type is used mainly for reaming pipe edges. The spiral type is best for smoothing the inside edges of pipe.

Note there are pipe cutters for both iron pipe and copper tubing.

If you have to cut pipe with a hacksaw, start as close as possible to the vise so that the free end of the pipe does not vibrate. Since saw cuts are rough, it will be necessary to "dress" the pipe after it has been cut.

Cutters for pipe are sized for ⅛" to 1", 1" to 3", 2" to 4", 4" to 6" diameters.

## TOOLS FOR COPPER TUBING

You will need a few special tools to work with copper tubing.

### Tube cutters (see "Cutting Pipe")

These are smaller and lighter than the type for iron pipe and a pipe vise is not ordinarily needed, since both tubing and cutter can be hand held. No cutting oils are required. Do not feed the cut too fast as this will have a tendency to flatten the tubing. Reaming will be needed after cutting, and filing, too, if outside burrs are present. On large-size tubing, hacksaws will work satisfactorily. Tube cutters are available in these dimensions: ⅛" to ¾", ⅛" to 1", ¼" to 1⅜" and ¾" to 2¼".

### Flaring tools

Many fittings to which tubing is attached are of the "flare" type. For these the tubing will have to be flared. To do this, first slip the sleeve nut on the tubing and use a flaring tool which will force the sides of the tube outwards.

### Tube benders

Normally, when it becomes necessary to change the direction of both tube and pipe, it will be easier to use angle fittings. If this is not practical, a tube or pipe bender will do the job. There are many types that will work satisfactorily, all the way from simple ones to those requiring motor drive. Unless you have considerable use for a tube or pipe bender, you should either rent the equipment or arrange to have the work done by an outsider.

### Blow torches

If you plan to do any extensive plumbing using large pipe or if you must melt quantities of lead, a well-designed blowtorch is a must. You will also require a melting pot and a cast iron ladle.

(In using a blowtorch, always follow the manufacturer's instructions carefully.)

### Propane torches

For the usual type of soldering job and for making solder connections in copper lines you will find a propane torch to be a sound investment. It will not only be much easier to handle but will be much safer, too.

### Cold chisels

These are indispensable if you have to cut cast iron pipe. You will also find much use for them in your general work, such as restoring damaged threads, cutting rusted pipe, and hammering out frozen plugs.

### Files

An assortment of files will be helpful for such jobs as smoothing the ends of pipe and tubing, repairing damaged threads and removing rust and dirt.

# 5

# PLUMBING QUALITY—
# IN PLANNING AND IN FIXTURES

If you are building or remodeling a home, the plumbing facilities should receive major attention. We recommend three steps when in the planning stage: first, make a list of the installations you would like, and bear in mind what your future needs might be; second, be sure a plumbing contractor checks your list and inspects existing plumbing, if it's a remodeling job; third, make a detailed cost estimate with your plumbing contractor's help. A higher first cost may be justified by long range economies in maintenance and repairs, and by greater enjoyment of your home.

A few tips: Place new fixtures close to existing pipes to minimize extra carpentry as well as pipe installation. Have as much work as possible done at the same time, while pipes in the walls are exposed. If you have future plans, install the basic connections now to avoid ripping open walls at a later date. Place up and downstairs plumbing on the same wall, and place kitchen and bath downstairs back to back to save unnecessary expense. Check local building codes carefully, since what you do must pass local government inspections.

## FIXTURES

A fixture is any appliance (tub, shower, toilet, sink, dishwasher, etc.) that uses or helps you use water. Those piping parts (faucets, traps, etc.) which are exposed to view—and, usually, are chrome plated—are considered part of the fixture (though they may not be furnished with the fixture), rather than parts of the water supply or drainage systems.

Select your fixtures with care, because your future satisfaction depends on your having selected those products that are best suited in style, performance and durability to your needs and desires.

Consider color. Colored fixtures cost only slightly more than white, and they represent a fine way to add decorative excitement to your facilities. Most manufacturers offer both pastel and bright colors and they are all attractive. It is important that all of your fixtures in a given color (even white) come from the same manufacturer, according to spokesmen from differing companies, such as Eljer and American Standard. Otherwise, you're likely to find that the colors don't really match from one fixture to the next.

A word about cost. As a general rule, the performance and appearance of plumbing fixtures and fittings are in proportion to their cost. The more expensive products usually perform better, last longer and are generally more attractive.

Do not make price alone your guide in selecting the proper fixtures and fittings. There is a multitude of designs available, across a wide range of prices. Each design offers unique characteristics and they warrant your consideration.

All plumbing fixtures and fittings from a reputable manufacturer are likely to be well-made; even economy products can be counted on to give years of trouble-free service. But the better grades will give you extra comfort, extra convenience, or a touch of elegance that is missing from the economy choice.

You must decide whether the added features are worth the added cost. But remember that the difference in cost between the least expensive and the most expensive is likely to be relatively small, and that it becomes almost unnoticeable when spread across the years of the average home mortgage.

*Toilets:* The toilet is the one fixture in your bathroom where design and quality control in manufacturer are of critical importance, and unlike the tub and lavatory which function primarily as basins to hold water, the toilet is a much more sophisticated mechanism (see Figure 5–1). It is semi-automatic in its function. Such things as proper relation between water volume and the interior design of the bowl must combine to create an efficient waste disposal system with automatic protection against sewer gases and unsanitary conditions.

Most residential toilets consist of a bowl and a tank. The function of the tank is to provide storage for sufficient water to create a proper flushing action. In commercial building, where large pipes are used for water lines, an adequate amount of water is provided directly from these supply lines, and all that's needed is a flush valve which shuts off au-

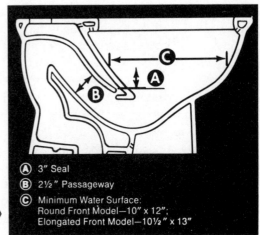

Ⓐ 3″ Seal
Ⓑ 2½″ Passageway
Ⓒ Minimum Water Surface:
Round Front Model—10″ x 12″;
Elongated Front Model—10½″ x 13″

Inside the ⟩
syphon jet bowl ⟩

**Figure 5-1**

tomatically after the proper amount of water has been provided. In residences, however, the water lines are too small to provide the amount of water needed, as fast as it is needed. Therefore, a toilet tank (or flush tank) is required.

What happens when you press the handle: (see Figure 1–11) **1.** When you depress the flush lever, it lifts a plug off an opening at the bottom of the tank. The water flows from the tank to the toilet, and provides the flush. **2.** In most tanks there is a large ball attached to a valve by a rod or arm. This ball floats on the surface of the water. As the water recedes, the ball drops, the valve opens, and water begins to flow into the tank. **3.** As the tank refills, the float ball rises, closing the valve and stopping the flow of incoming water at the proper level within the tank to provide an efficient flushing action with a minimum amount of water. This level is indicated by a mark inside the tank. The water level can be adjusted to reach this level by bending the float arm. You may be tempted to set the water level below this line to save water. Don't do it. With the reduced volume of water, the flush is likely to incomplete and a second flush may be required to properly clense the bowl. You'll wind up wasting water instead of conserving it. **4.** A stand pipe (overflow) prevents the tank from overflowing if the float valve fails to shut off properly. Running from the float valve to this overflow outlet is a small tube. It is important that this tube be placed so that water discharged from it will go directly into the overflow pipe. This water replaces the water level in the bowl after flushing. The toilet will not flush properly if this level is too low.

Not all toilets are alike. All toilets remove waste, but some work better than others. A good toilet should operate quietly and should provide a large water surface area to prevent contamination and fouling. There are four basic kinds of residential toilets, distinguished by their different flushing actions.

*Washdown*: The washdown toilet is the least expensive, the least efficient and the noisiest. It's flushed by a simple wash-out action, and may clog more easily than the types to be mentioned. Much of the bowl area is not covered by water and is subject to fouling, staining and contamination.

*Reverse trap*: The reverse trap is the least expensive of the siphon action toilets. It is flushed by creating a siphon action in the trapway, assisted by a water jet located at the inlet to the trapway. This siphon pulls the waste from the bowl. It is moderately noisy, but efficient. The trapway must pass a 1½" ball and the siphon makes this closet less likely to clog. More of the interior bowl surface is covered with water than in the washdown models. Therefore, it is less subject to fouling. It is only slightly more expensive than the washdown and is a better buy.

*Siphon jet*: Siphon jet bowls are improved versions of the reverse trap bowls. They have a larger water surface, with most of the interior surface of the bowl covered with water. The trapway is larger and must pass a 2⅛" ball, thus the flushing action is quieter and less subject to clogging. Siphon jets are usually more expensive than reverse trap toilets.

*Low profile*. "Siphon action" is a term applied by some manufacturers to the flushing action of low silhouette, one-piece toilets. These toilets are elegant but more expensive than a siphon jet toilet. The trapway must pass a 2" ball. However, if you want an almost silent flushing action, almost no dry surfaces on the bowl interior, and a low profile extremely attractive toilet, you may feel the extra cost is well spent.

Styling—elongated versus the round bowl. Most toilets are available with either round or elongated bowl rims. The elongated bowl (sometimes referred to as "extended rim") is 2" longer from the front edge of the rim to the back of the toilet. You will probably be most satisfied with an elongated bowl. It is more comfortable, more attractive, and, since it provides a larger interior water surface, more sanitary and easier to keep clean. And it costs very little more than a round bowl (see Figure 5–2).

*On-the-floor toilets*. These attractive and functional toilets, which were first developed for application in commercial and public washrooms because they facilitate floor cleaning and maintenance, have been redesigned by many manufacturers to meet the esthetic requirements of resi-

**Figure 5-2**

dential use and to permit use with flush tanks. They offer the housewife the same easy floor cleaning benefits that made them so acceptable in commercial applications. They are quite attractive, quite efficient, but are also expensive to install since they require a special metal "chair carrier" installed inside the wall to support them. This special construction within the wall usually requires that the wall be of 2″ × 6″ studding instead of 2″ × 4″s.

Some studies indicate, however, that if the installation of this type of toilet is planned for at the outset in the new home construction, its final cost may be comparable with that of a conventional toilet of similar quality.

*Special-use toilets.* A number of special designs in toilets are available for special uses. A toilet with a triangular tank may be well worth its extra cost because of it's unique use of floor space—it takes the greatest advantage of corner space—floor area normally wasted. This toilet, with adjacent grab bars, is ideal if there are invalids or aged in your family. Special high toilets are for use by infirm individuals, also included in the lines of most major manufacturers.

*The bidet.* The bidet is perhaps the most misunderstood bathroom

**Figure 5-3**

fixture. This fixture is not designed for internal feminine hygiene; it is designed for use by the entire family, to wash the perineal area after using the toilet. For convenience, the bidet should be installed beside the toilet. It is unique in appearance, something like a toilet (see Figure 5–3).

A pop-up stopper is provided for rinsing. And a flushing rim serves to rinse the entire inner surface of the bowl after using.

The user sits on the fixture, facing the wall and the hot and cold water controls. A transfer valve first produces a soothing, comfortable, rinsing spray, then diverts the water to the flushing rim to clean the bowl.

Of widespread use in Europe and South America, the bidet has only, in recent years, become recognized in the United States as a logical twin to the toilet and a helpful means of personal hygiene for the entire family.

If you want a truly modern bathroom, you may wish to investigate its purpose, advantages and costs more thoroughly.

*The bathtub.* Bathtubs come in a variety of sizes and shapes. The length of a standard rectangular tub ranges from 4′ to 6′ and its height ranges from 12″ to 16″ (see Figure 5–4). The height of the tub is normally computed from the floor to the top of the rim; the actual depth of water the tub will provide, however, is governed by the distance from the bottom of

**Figure 5-4**

*Marlite*® *Product*

the tub to the overflow outlet, a dimension not always printed in literature and sometimes obtainable only by measuring a sample tub in a dealer's showroom. Nevertheless, regardless of water depth inside the tub, the higher the tub, the less water you'll have splashing over the sides onto the floor.

Make the tub long and deep. If you like a relaxing, soaking bath, you'll probably want a long, wide and deep tub with a sloping end to relax against. If so, make certain your builder knows what you want because the width of many bathrooms is determined by the length of the tub. If the builder frames your bathroom for a five-foot tub, there's no way you're going to get a six-foot tub into the space.

Types of bathtubs—rectangular. The most generally installed tub is rectangular and is recessed into a niche, which is then tiled for showering. Corner tubs are for installations where the back and only one end of the tub will be against walls. This model is less popular and more expensive.

*Square bathtub.* If you want to create a unique looking bathroom, consider a large square tub or one of the unique shapes on the market, all of which are premium priced. Small square tubs work well in areas where space is a problem, although they are not large enough for adult tub bathing. They are ideal for a children's bath or as a luxury shower base.

*Shower stalls.* A shower stall offers an alternate for those who prefer a shower to a soak. Make it large enough. A 36″ × 36″ shower stall is the minimum adequate size for adult showering. The larger the stall, the more comfortable it will be to use.

*Bathtub materials.* Most bathtubs are manufactured from one of three materials; molded cast iron with a porcelain enamel surface, formed steel with a porcelain enamel surface, or a molded del-coated glass fiber reinforced polyester resin. The polyester resin units are generally known as fiberglass tubs.

*Cast iron.* Cast iron tubs were first manufactured in 1870. The enameling process used dates back even further, to ancient Egyptians and Assyrians. These tubs are available in 4′, 4½′, 5′, 5½′, and 6′ lengths. They range in width from 30″ to 48″, and in depth from 12″ to 16″. The combination of thick, glossy porcelain (approximately $1/16''$) and the heavy rigid cast iron body makes these tubs less susceptible to damage than tubs made of other materials. They are heavy. A 5′ cast iron tub will weigh approximately 300 lbs. Some sizes weigh as much as 500 lbs.

A short front apron is available on some cast iron tubs to facilitate recessing into the floor for a sunken tub application. Sunken tubs constitute a safety hazard, and precautions should be taken to provide numerous grab bars for safety.

*Formed steel.* Formed steel tubs with a porcelain enamel finish were developed to provide a lightweight (about 100 lbs.) tub that would be less expensive than cast iron. They are ideally suited for upper story installations or for remodeling because of the ease with which they can be moved into place. Steel tubs are usually available in two lengths only: 4½′ and 5′. Widths range from 30″ to 31″ and the typical depth is 15″ to 15½″.

If available in your area, a steel tub formed from one piece of steel is preferred. Some tubs are produced in two pieces: the apron being formed separately and welded into place prior to enameling. The welded seam may be unattractive. You may wish to request a sound-deadening undercoating on your steel tub. Most manufacturers offer such a coating as an option. Cost is normally reasonable and may be justified by the reduction of shower noise.

*Fiberglass.* Fiberglass bathtubs are a recent development. They have been in widespread use since about 1968. Those manufactured by reputable plumbing fixture manufacturers can be bought with confidence. Be very careful and select only nationally advertised brand names.

One-piece fiberglass tub-showers solve the problem of maintaining a watertight, nice looking joint where the surrounding walls and the bathtub come together. These are combination bathtubs and wall surrounds, providing walls up to about 73″ above the floor in one piece. Recently there have been numerous variations in design and size for special applications, but generally those used in residences are of the rectangular combination tub and wall surround type, and about the only length readily available nationwide is 5′.

The surface of a fiberglass tub is provided by a gel coat. This is smooth and attractive, but cannot be made as durable as porcelain.

*Shower stalls.* Shower stalls are available in enameled steel or fiberglass, or may be assembled on site using a tile and various types of waterproof bases. Like fiberglass tub-showers, shower units of these materials are molded in one piece. Soap dishes or seats may be molded-in. And, with one-piece construction, there is no danger of leakage.

*Maintaining bathtub beauty regardless of materials.* The porcelain on cast iron and steel tubs gives a hard, glasslike surface like fine dinnerware. The surface of fiberglass units is a gel coat, as stated—a strong polyester resin that is bright and extremely easy to clean. Abrasive cleaners will harm these finishes and should not be used. The dulling which results from the use of abrasive cleaners on fiberglass units will also occur when such cleaners are used on porcelain. It just takes longer.

### The lavatory

No other bathroom fixture comes in as many styles, sizes and shapes as the lavatory. Because it is often surrounded by attractive light fixtures, medicine cabinet, countertop, vanity, etc., the area it occupies becomes the focal point of interest in the average bathroom. This frequently causes a great preoccupation with decor in lavatory selection at the sacrifice of practical consideration of the multitude of uses which this area must perform (see Figure 5–5).

With such a wide variety of choices available, it is possible to select lavatories that fit the needs of your family and still satisfy your decorating impulses. With this in mind, give some thought to the space you have available and the functions for which the lavatory is used. Consider that the lavatory may be needed for washing delicate garments, in addition to washing face and hands and shampooing. On this basis, a small lavatory should be used only when space is a problem.

**Figure 5-5**

## FIVE LAVATORY STYLES

1. *Flush-mount*. This type of lavatory requires a metal ring or frame to hold it in place. This is a popular and inexpensive style but does cause a cleaning problem at the junctures of the rim with the countertop and the lavatory.

2. *Self-rimming*. This style is quite attractive since it requires no metal framing ring. Lavatories in this style project above the countertop and are designed so that the rim of the lavatory rests on the countertop. Properly sealed, this type offers easy cleaning.

3. *Under-the-counter*. This type of lavatory mounts beneath a finished opening cut in the countertop, which is usually marble or a synthetic material that looks like marble. With this lavatory, the fittings are mounted through the countertop. Although the total effect is quite handsome and luxurious in appearance, the seam where the lavatory meets the underside of the countertop is prone to collect dirt and is difficult to clean.

4. *Integral lavatory and counter*. Recently, a number of plastic or "synthetic marble" countertops with an integral basin have been introduced. If you are attracted by their clean, seamless lines,

make certain the design you are considering is supplied with an adequate overflow.

5. *Wall-hung*. The conventional wall hung lavatory is most often rectangular in shape. Other shapes and special corner units are also available. It works well where space is scarce, or where a storage space is not required and the homeowner is not concerned about the exposure of the plumbing beneath the bow.

*Lavatory materials*. Most lavatories are made of cast iron, vitreous china, or formed steel. The casting or moulding methods employed in making lavatories of cast iron and china permit great license in design and shape. The punching or pressing process used to make steel lavatories limits the form of the final product. However, in a flush-mount or under-the-counter lavatory, it is difficult for the casual observer to determine the difference between these three materials.

## YOUR LAUNDRY ROOM

The stationary laundry tub is a plumbing fixture, but one you may not think about when selecting the fixtures for your home. You will find that laundry tubs come in a wide range of materials and sizes. Some are difficult to clean, some may not hold up under constant exposure to some of the harsh bleaches and chemicals used in modern washday activities. We suggest you consider the enamel cast iron tubs, available from many makers of plumbing fixtures, and especially designed for laundry use. They are easily cleaned, and the porcelain finish holds up well for most laundry needs. These tubs come in single compartment and double compartment styles—and there are even styles available to install in a counter-top. Take a look before you buy.

## BATHROOM FITTINGS

Fittings, faucets, spigots—regardless of what you call them, they're the working parts of your bathroom. Plumbing fittings (including showerheads and pop-up drains) are mechanical parts, subject to wear. They do a lot of work, and they wear out, and they will need to be repaired or replaced long before anything else in your plumbing system needs maintenance (see Figure 5–6).

Bathroom faucets are designed to mix hot and cold water. A faucet falls into one of four classifications, depending on how it performs this function.

## TWO HANDLE (BLADE) AND (DURALAC) LAVATORY FAUCETS

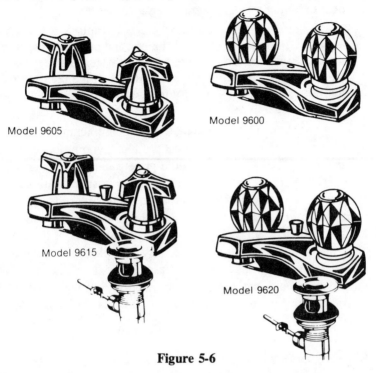

Model 9605

Model 9600

Model 9615

Model 9620

**Figure 5-6**

1. *Two-valve*. This is the most common bathroom fitting. Each of its two handles operates a valve; they are turned individually, adjusting the hot and cold water to provide the desired temperature and volume from the spout. Special wrist-action handles are available to permit usage by arthritic individuals or others who cannot grip ordinary handles.

2. *Single control*. The single control valve may be operated by either a knob or a lever. By moving the control to the right or left, you control the water temperature. Volume is controlled by moving the same control in and out or backwards and forwards.

3. *Pressure balancing*. Volume on the pressure balancing valve is usually preset. The user sets a temperature, and the valve maintains that temperature automatically. This is accomplished by a pressure-sensing device, which automatically decreases the flow of either the hot or the cold water when the operating pressure in the opposite line drops. Should the pressure of the cold water supply drop sufficiently to cause scalding, the valve automati-

cally shuts off the hot water flow. The pressure balancing valve is most commonly found controlling a showerhead.

4. *Thermostatic*. The thermostatic control employs a heat-sensing device to automatically adjust the hot and cold volume to maintain a pre-selected temperature of mixed water. This type of valve gives more precise temperature control than the pressure balancing valve, and it usually permits the user to control volume as well as temperature.

*The cost of fittings*. Most major fixture manufacturers produce fittings in three price ranges. As with most products, you probably get most for your money by buying in the middle price range. This is usually the line designed to meet architectural and engineering specifications for use in public and commercial buildings and, therefore, normally offers maximum durability and service life with the lowest maintenance.

Many manufacturers incorporate the working parts of these fittings with elaborate designs and exotic materials (gold plating, special decoration, onyx handles) to produce luxury brass. These decorations and materials increase cost and may not improve function, but the decorator look may warrant your paying the increased cost for luxury brass.

## A LITTLE LOVING CARE

Here are some things you can do to make your fittings last longer. When you operate a two-valve fitting, do not close the valve tightly. Turn it just far enough to stop the flow of water. If you follow this operating principle, you should get several years of service without having to replace the washers in the valve. If you consistently turn them off as tightly as you can, you may have problems in a few months.

Chrome is the most durable metal surface available. However, even a chrome finish should not be cleaned with an abrasive cleaner or one with acid as one of the ingredients. If you like and want polished brass, be prepared to clean it once a week with brass cleaner.

Gold is a very soft finish and the plating, when used on plumbing fittings, is generally very thin. Do not clean with anything other than a soft cloth and water. Expect to have the gold plating replaced in a few years.

Antique finishes are protected with lacquers and will have to be refinished in a few years.

A word about warranties. According to the experts at Eljer, one of the major quality manufacturers of plumbing fixtures, most makers guarantee plumbing fixtures against manufacturing defects for a period of one year after installation and they limit their liability to providing a replacement product of like size and quality, but do not pay for removal and installation. Generally, if the product is satisfactory when it is installed no latent defects will show up. Don't be unreasonable. But, if you find anything you think is wrong, contact your plumber and you will most likely get a fair assessment of the responsibility. Your best bet is to carefully inspect all products before installation and reject any that don't suit you.

# 6

# WORKING WITH PIPE AND TUBING

Pipe is available in many kinds of materials: steel, wrought iron, concrete, clay, copper which is either rigid or soft tubing, plastic, fiber, and asbestos-cement. The steel and iron types are used for steam, gas and water lines and sold in sizes from ⅛″ to 6″ diameter. Though it is cheaper than copper, it cost more in time to install it because of the need for threading operations. Wherever pipe lines are subject to impact, steel and wrought iron are used.

Copper which is rigid can be threaded or soldered, and is lighter and of better appearance than steel. Soft copper tubing is easiest to install, costs less because it is thinner, and since it impedes the flow of water less than steel, a smaller diameter line can be used when replacing steel, often. See Figure 6–1 for a quick summary of pipe information.

Plastic pipe is the easiest to install, but codes should be carefully checked, since they may not be allowed for some purposes; the technology in this industry is moving faster than the others, because of the material's relative newness. The clay, fiber, concrete and cement types are used for drainage.

## IDENTIFICATION OF FITTINGS

It is important to note that all pipe sizes refer to the inside diameter (I.D.) of pipe. Fittings are identified according to the size of pipe they fit. If pipe is unthreaded, all lengths are referred to simply as pipes; but with threaded pipe the short lengths (up to 12″) are called nipples. Pipes are always assembled by fitting their ends inside of the fittings used to join them—hence the terminology "male" (pipe ends) and "female" (fitting openings). However, only coupling-type fittings have both or all female

PIPE DATA AT A GLANCE

| TYPE OF PIPE | EASE OF WORKING | WATER FLOW EFFICIENCY FACTOR | TYPE OF FITTINGS NEEDED | MANNER USUALLY STOCKED | LIFE EXPECTANCY | PRINCIPAL USES | REMARKS |
|---|---|---|---|---|---|---|---|
| Brass, Threaded | No threading required. Cuts easily, but can't be bent. Measuring a job rather difficult | Highly efficient because of low friction | Screw on Connections | 12 ft. rigid lengths. Cut to size wanted. | Lasts life of building | Generally for commercial construction | Required in some cities where water is extremely corrosive. Often smaller diameter will suffice because of low friction coefficient |
| Copper-Hard | Easier to work with than brass | Same as brass | Screw on or or Solder Connections | 12 ft. rigid lengths. Cut to size wanted. | Same as brass | Same as brass | |
| Copper-Soft | Easier to work with than brass or hard copper because it bends readily by using a bending tool. Measuring a job not too difficult | Same as brass | Solder Connections | Coils - usually soft | Same as brass | Widely used in residential installations | |
| Copper Tubing, Flexible | Easier than soft copper because it can be bent without a tool. Measuring jobs is easy. | Highest of all metals since there are no nipples, unions, or elbows | Solder or Compression Connections | 3 wall thicknesses: 'K'-Thickest 'L'-Medium 'M'-Thinnest 20 ft. lengths or 15 ft., 30 ft., or 60 ft. coils (Except 'M') | Same as brass | 'K' is used in municipal and commercial construction. 'L' is used for residential water lines. 'M' is for light domestic lines only - check Code before using. | Probably the most popular pipe today. Often a smaller diameter will suffice because of low friction coefficient. |
| Wrought Iron (or galvanized) | Has to be threaded. More difficult to cut. Measurements for jobs must be exact. | Lower than copper because nipples unions reduce water flow. | Screw on Connections | Rigid lengths, up to 22 ft. Usually cut to size wanted. | Corrodes in alkaline water more than others. Produces rust stains. | Generally found in older homes | Recommended if lines are in a location subject to impact. |
| Plastic Pipe | Can be cut with saw or knife. | Same as copper tubing | Insert couplings, clamps; also by cement. Threaded & compression fittings can be used (Thread same as for metal pipe) | Rigid, semi-rigid & flexible. Coils of 100-400 ft. | Long life & it is rust & corrosion-proof. | For cold water installations. Used for well casings, septic tank lines, sprinkler systems. Check Codes before installing. | Lightest of all, weighs about 1/8 of metal pipe. Does not burst in below freezing weather. |

**Figure 6-1**

sides. Bushing-type fittings, also called street fittings, are male at one or more sides and are used to join a pipe with another fitting instead of another pipe.

The standard coupling-type fittings consist of couplings, ells (also elbows or bends), tees, wyes, crosses and unions (the latter being a coupling which can be separated into two parts without disassembling the pipe run it is in). See Figures 6-2, 3, and 4 for illustrated information on these points. An ordinary fitting of this type simply joins two or more pipes which are identical in size and also the same type (threaded, copper, etc.). If the pipes are the same type but are of different sizes, a reducing fitting is required, such as a reducing coupling (see Figure 6-2). When the pipes are the same size but of different materials (copper and steel), an

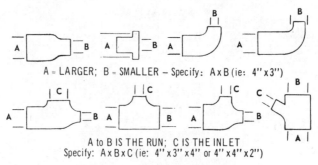

A = LARGER;  B = SMALLER – Specify:  A x B (ie:  4" x 3")

A to B IS THE RUN;  C IS THE INLET
Specify:  A x B x C (ie:  4" x 3" x 4" or 4" x 4" x 2")

**FITTING TYPES AND DATA**

**Figure 6-2**

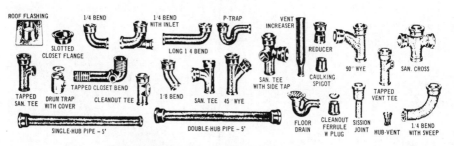

TYPICAL HUB-AND-SPIGOT CAST-IRON PIPE AND FITTINGS

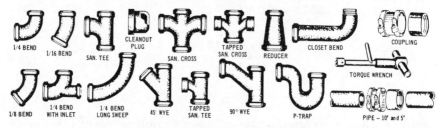

TYPICAL NO-HUB CAST-IRON PIPE AND FITTINGS

**Figure 6-3**

adapter is used to join them. There are also specialized reducing-adapting types of fittings which combine both features, but these are seldom used.

Very seldom is it absolutely necessary (because of space limitations) to use any special fittings other than reducers and adapters. With these two and standard fittings you can assemble any pipe run, even though it may change in size and/or type, by using two fittings from this group instead of one special fitting, whenever necessary. For this reason the available fittings shown here do not include very many special types.

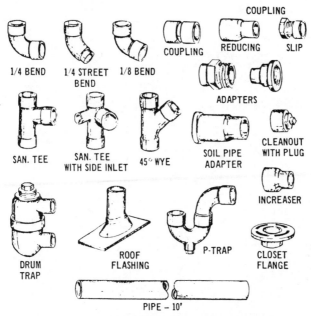

**TYPICAL COPPER PIPE AND FITTINGS**

**Figure 6-4**

### TYPES OF DRAINAGE PIPES

A cast-iron and steel system: in this system cast-iron pipes and fittings (see Figure 6–3), which are available in 2″, 3″ and 4″ sizes with pipe in 5′ and 10′ lengths are used for all main stacks, for any toilet branch drain, for any other branch drain that is buried in the ground or in concrete, and for the house drain. Galvanized steel pipe and fittings (available in 1½″ and 2″ sizes with pipe in 10′ lengths) are used for all other parts of the system, such as a secondary stack, vent lines which join a stack and branch drains other than those mentioned above. Special sanitary-type drainage fittings designed to avoid accumulation of solids must be used in all parts of a system carrying waste water—standard fittings are used in those vent parts where gas alone will be carried. Consequently, all cast-iron fittings are the sanitary type; but fittings for steel are available both in sanitary and standard type.

Hub-and-spigot type cast-iron pipe and fittings are joined by inserting a spigot end into a hub and caulking the joint. The new no-hub cast-iron pipe and fittings are more easily joined by the use of special couplings (which require no skill to assemble). Steel pipe and fittings are joined by screwing a male-threaded end into a female-threaded fitting.

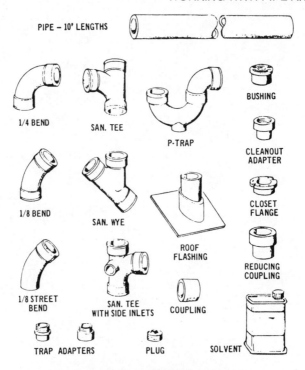

PIPE – 10' LENGTHS

1/4 BEND

SAN. TEE

P-TRAP

BUSHING

CLEANOUT ADAPTER

1/8 BEND

SAN. WYE

CLOSET FLANGE

ROOF FLASHING

REDUCING COUPLING

1/8 STREET BEND

SAN. TEE WITH SIDE INLETS

COUPLING

TRAP ADAPTERS

PLUG

SOLVENT

**TYPICAL PLASTIC PIPE AND FITTINGS**

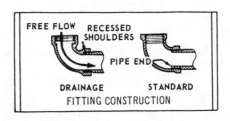

FREE FLOW   RECESSED SHOULDERS

PIPE END

DRAINAGE         STANDARD

FITTING CONSTRUCTION

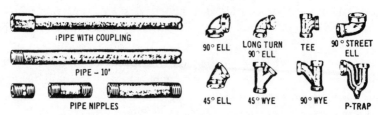

PIPE WITH COUPLING

PIPE – 10'

PIPE NIPPLES

90° ELL

LONG TURN 90° ELL

TEE

90° STREET ELL

45° ELL

45° WYE

90° WYE

P-TRAP

**TYPICAL GALVANIZED STEEL PIPE AND DRAINAGE FITTINGS**

**Figure 6-5**

The words "tap" or "tapped" used in identifying a cast-iron fitting (see Figure 6–5) mean that the fitting has a threaded opening for steel pipe. Standard steel-pipe fittings are shown further on in this chapter. Cast-iron pipe, fiber or tile pipe in 4″ sizes is used for the house sewer, with an adapter to connect with the house drain, if needed.

### A COPPER-PIPE SYSTEM

In this system copper pipe (with soldered joints) is used for all parts which are not buried underground or in concrete. Pipes and fittings are available in 3″ size for all stacks, toilet branch drains and suspended (above ground) house drains, and in 1½″ size for all other above ground drain and vent parts. See Figure 6–4. Pipes are in 10′ lengths. Cast-iron is used for any part of the system that is below ground or in concrete—and a special fitting accomplishes the junction. Cast-iron, fiber or tile pipe in 4″ size is used for the house sewer—again with a special fitting as needed.

### A PLASTIC PIPE SYSTEM

In this system plastic pipe (with solvent-cemented joints) is used throughout for all above-ground or buried parts, out to the house sewer. Pipes and fittings are available in 3″ size for main stacks, toilet branch drains and house drains—in 2″ size for secondary stacks and some branch drains (a floor drain, for example)—and in 1½″ size for all other branch drains and vent runs. Pipes are in 10′ lengths. All fittings are specially designed for either drainage or venting use. (See Figure 6–5) Special fittings are used for conversion where house drain joins house sewer, or where in an old house it is necessary to join with cast iron or steel pipe. If your code permits, plastic pipe can also be used for the house sewer.

### TYPES OF WATER PIPES

The system parts shown here are for a complete water piping system, with these exceptions: exposed fixture supply lines, which should be ordered with each fixture; special fittings required to connect the house main with the building main, or the latter with the water source; special fittings to connect the system to water using appliances, which should be ordered with the appliance. If your water source is a utility main, the utility generally makes the connection. If it's a private source you may be responsible.

In a galvanized steel pipe system, furnished in 10′ lengths, steel pipe and galvanized malleable-iron fittings are used throughout, within the building. Pipes and fittings are available in ⅜″, ½″, ¾″, 1″ and larger

THE FITTINGS SHOWN BELOW ARE USED FOR
ASSEMBLING WATER LINES AND IN THE VENT PORTION
ONLY OF DRAINAGE PIPE RUNS

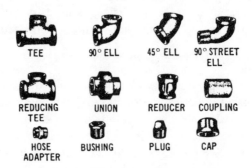

**TYPICAL STANDARD STEEL-PIPE FITTINGS**

**Figure 6-6**

sizes, but generally only the ½" and ¾" sizes are used in home plumbing. All parts are joined by screwing them together. Valves and sill faucets are not classified as fittings. These, generally, are made of brass. Steel pipe should not be buried in the ground under a building or in concrete. It can be used underground out from under a building where it can be reached for repair when necessary. See Figure 6–6 for typical standard steel-pipe fittings.

### A COPPER-TUBING SYSTEM

In this system rigid copper tubing (type L) copper fittings and copper valves are used for all parts within the building. Pipes, etc., are available in ⅜", ½", ¾" and 1" and larger sizes, but again only the ½" and ¾" are commonly used in home plumbing. Pipe is furnished in 10′ lengths. All parts are soldered (sweated) together. Type K (heavy) flexible copper tubing is solder-joined to the house main, and is used for the building main. (See Figure 6–7)

Though rigid copper is preferred for new home use or wherever piping is exposed, flexible tubing is often easier to feed through finished partitions when doing remodeling work. It also is advantageous for such special applications as connecting a furnace humidifier, or even a water heater, water softener, laundry tub, etc., where it would be difficult to install the number of fittings that would be required if rigid tubing were used. Flexible tubing (type L or K) is available in 30′ or 60′ coils. The same sweat fittings used with rigid tubing are commonly used, but flare-

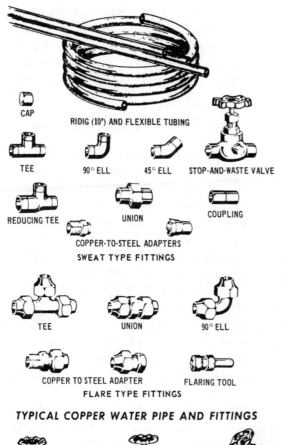

CAP

RIDIG (10') AND FLEXIBLE TUBING

TEE 90° ELL 45° ELL STOP-AND-WASTE VALVE

REDUCING TEE UNION COUPLING

COPPER-TO-STEEL ADAPTERS

SWEAT TYPE FITTINGS

TEE UNION 90° ELL

COPPER TO STEEL ADAPTER FLARING TOOL

FLARE TYPE FITTINGS

## TYPICAL COPPER WATER PIPE AND FITTINGS

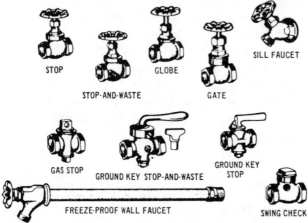

STOP GATE SILL FAUCET

STOP-AND-WASTE GLOBE

GAS STOP GROUND KEY STOP-AND-WASTE GROUND KEY STOP

FREEZE-PROOF WALL FAUCET SWING CHECK

## TYPES OF VALVES USED IN WATER-SUPPLY SYSTEMS

**Figure 6-7**

type fittings can be used (if your code permits), and generally are for underground connections of type K tubing. Flare-type fittings have fitted slip nuts which are tightened against tube ends (it must first be flared) to compress it against a seat on the fitting. Once assembled a connection seldom can be taken apart and reassembled successfully unless a new nut is used.

### TYPES OF FIXTURE FITTINGS

In modern bathrooms and kitchens, durable, easy to clean chrome-plated fittings are preferred for all parts of drainage and water-supply systems which will be exposed to view (see Chapter 5). Faucets, shower heads, strainers and similar parts that are styled for a particular fixture generally are available with each fixture (if it is ordered "with fittings"), but in all other cases the fittings must be listed and ordered separately.

For draining a sink you need a tail piece (which fastens to the bottom of the sink strainer) and a trap which connects the tailpiece to the plumbing system outlet in the wall or floor. For a lavatory, to review, you need a trap and if the outlet is in the wall and less than 18″ from the floor, also a drain extension which will serve as a tailpiece (see Figure 6–8). A laundry tub will require sink parts if it has a strainer, or lavatory parts if it has a drain fitting. Sink parts are 1½″ size; lavatory parts are 1¼″ size. All parts are joined together by slip joints, which use replaceable rubber or neoprene washers.

For a toilet you need one supply pipe; for a sink, lavatory or tub you need two. Pipes are of various lengths (12 to 20″) and come in ⅜″ and ½″ sizes. They are bendable copper with or without chrome plating, and generally have a fitting at each end—one end to fit faucet or the toilet ballcock, the other to fit the water-supply nipple in the wall or floor. The easiest way to order is to specify the type of fixture and whether to an in-wall or in-floor nipple. Except for a sink you also have a choice between a supply pipe with a shut-off valve or without. Sinks and tubs generally take ½″ size pipe; toilets and lavatories ⅜″. For ⅜″ size use a ½″ × ⅜″ bushing at the ½″ nipple end. Botch connect to all standard faucets.

### HOW TO ASSEMBLE PIPES AND FITTINGS—CAST IRON

With cast-iron pipe it is important to note that the flow through it and any fitting must always be in at the hub end and out at a spigot one. This goes for all types of pipe.

Use of double-hub pipe: Where shorter lengths are needed, the 5′ standard sections must be cut off as required. If a double-hub length is cut

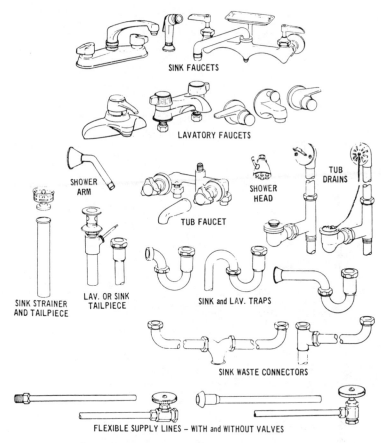

SINK FAUCETS

LAVATORY FAUCETS

SHOWER ARM

SHOWER HEAD

TUB DRAINS

TUB FAUCET

SINK STRAINER AND TAILPIECE

LAV. OR SINK TAILPIECE

SINK and LAV. TRAPS

SINK WASTE CONNECTORS

FLEXIBLE SUPPLY LINES – WITH and WITHOUT VALVES

**TYPICAL FIXTURE WASTE AND SUPPLY FITTINGS AND TRIM**

**Figure 6-8**

in two each end will have a hub, thus making both ends usable, and reducing the amount of waste. (See Figure 6–9.)

*Measuring pipe lengths*: The spigot end of one pipe length must fit all the way into the hub of the adjoining pipe or fitting, when assembled. Make these allowances for hub depth: 2″ pipe hub depth is 2½″, 3″ pipe hub depth is 2¾″, 4″ pipe hub depth is 3″. Figure 6–9 shows how to make measurements prior to cutting off a pipe. Note that the pipe is measured from the inside shoulder of the hub outward to the distance required. A hacksaw is excellent for making the mark. If necessary, the spigot end of a fitting (such as a long 90° ell or closet bend) can be measured for cutting off in much the same way.

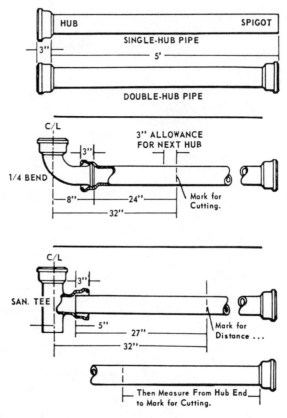

**MARKING A 4" PIPE FOR A 32" DISTANCE**

**Figure 6-9**

*Cutting pipe*: Using a hacksaw or file circle the pipe at your mark with a groove about $^1/_{16}''$ deep. Be sure to make the groove straight around and square with the pipe. Lay the pipe on a board (see Figure 6–10) as shown with the groove at the top edge of the board. Use a cold chisel and hammer to tap in the groove. Hammer lightly as you circle the pipe once, then continue to circle the pipe repeatedly, striking harder and harder blows until the pipe breaks.

*Vertical joints*: For a vertical joint a spigot end is always inserted downward into a hub end. (See Figure 6–11) Seat the spigot squarely in the hub so that pipes are aligned straight, with the spigot centered in the hub. Support the pipes as required to hold them steady. Pack the space between the spigot and hub, evenly all around, with oakum—to a depth of

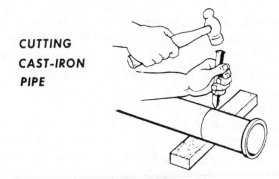

**CUTTING CAST-IRON PIPE**

**Figure 6-10**

## Making Vertical Joints

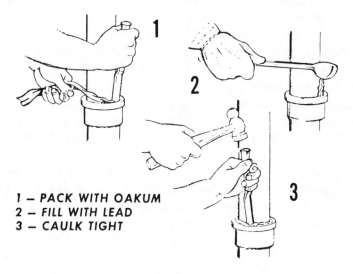

1 — PACK WITH OAKUM
2 — FILL WITH LEAD
3 — CAULK TIGHT

**Figure 6-11**

about ¾" to 1" from the top of the hub. The oakum, furnished in strands about ½" thick, is packed in with a yarning iron as in Figure 6–11. It must be firmly packed (if properly done it will hold the top pipe upright without a support).

Now fill the remaining space—up to the top of the hub—with molten lead, melted beforehand in a plumber's furnace. It should take about 1½ lbs. of lead to fill a 2" pipe joint, 2½ lbs. for a 3" pipe, and 3½ lbs. for a 4" pipe joint. Lead should not be red hot when pouring—this would burn the oakum—but it should be hot enough to flow all around the joint and fill it uniformly without the formation of bubbles or lumps. Several minutes of heating past the melting point will usually accomplish the proper condition. If lead is just right you can usually fill a joint with one pouring, and the lead will begin to set almost immediately thereafter. *Note*: always heat the ladle over the furnace flame prior to dippling lead. A cold ladle will spoil the pouring.

The lead shrinks as it cools. To keep the joint tight it must thereafter be caulked. Use a caulking iron and hammer. First work around the joint several times to spread the inner half of the lead rightly against the spigot. Then repeat to the outer half against the hub. Caulk it tightly and uniformly but do not hammer excessively hard as this could crack the hub.

*Making horizontal joints*: A horizontal joint is more difficult than a vertical one. If you have a horizontal run that requires several joints it will be easier to preassemble as many of the pieces as possible vertically—then lay the assembled pieces down as a unit to make the remaining horizontal joints needed.

The only difference between a horizontal joint and a vertical one is that a joint runner must be used to prevent the molten lead from spilling out as it is poured (see Figure 6–12). The joint runner is a pliable device that can be clamped around the spigot-end pipe up against the hub to seal in the area to be filled with lead. Clamp the runner snugly around the pipe then slide it (with hammer taps) tightly over against the hub top—all around except right at the top where there will remain a triangular opening (between the two runner ends and the hub) into which the lead can be poured. Pour in lead until its level rises to the top of this opening. If the pouring results in an excess of lead at any point, cut this off with a cold chisel prior to caulking.

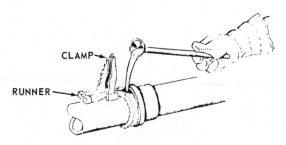

**Figure 6-12**

*Note*: lead wool, although slightly more expensive, may be easier to use than molten lead. If substituted, it is twisted into a strand like the oakum, laid in between spigot and hub, and caulked in the same manner as poured lead. Care must be taken to make the strand thick enough (about half again as thick as the space to be filled) and uniform throughout (no thin or lumpy spots). Caulking will mash and spread a twisted strand to form a tight seal—but it will not bind together two separate strands (that aren't already twisted together) firmly enough to keep them from later separating. In short, you can't join cold lead to cold lead by hammering.

*No-hub cast-iron pipe*: It is important to note that there is no required direction of flow through this pipe, but there is through any sanitary fitting used with the pipe. Be sure to face the fittings correctly. In measuring pipe lengths, the ends of two pipe lengths—or a pipe and a fitting—butt squarely together inside of the sleeve coupling that joins them (see Figure 6–13), separated only by the small shoulder molded on the inside of each coupling. Therefore, for all practical purposes, no allowance need be made for overlap or joining space when measuring pipe lengths. When measuring for a straight run simply add together the end-to-end measurements of the pipes and fittings which will comprise the run. If measuring for a run that will make a bend or have a branch, do the same—except measure only to the center of the bend or tee involved.

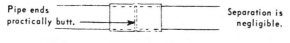

Pipe ends practically butt.                              Separation is negligible.

**Figure 6-13**

*Cutting*: This pipe is cut in the same manner as hub-and-spigot cast-iron pipe.

*Assembling*: See Figure 6–14. Place the sleeve coupling over the end of one pipe (or fitting) as far as it will go—and put the stainless-steel shield (with band clamps) around the other pipe end. Next, push the second pipe end into the sleeve coupling as far as it will go—so that the two pipe ends are now firmly butted against the integrally molded shoulder inside of the coupling. Finally, slide the shield to center it over the coupling, then tighten the two band clamps.

For proper tightening of the band clamps a torque wrench that automatically releases at 60″ lbs. is required. This allows the joint to set and still retain 48″ lbs. of pressure tightness as required by manufacturer's specs. Either a hand torque wrench or an air impact wrench may be used.

### Assembling No-Hub Pipe

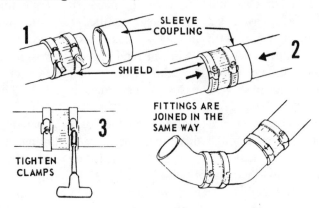

**Figure 6-14**

## STEEL PIPE

It is again important to note that there is no required direction of flow through a steel pipe or a standard fitting—but there is through any sanitary fitting. Be sure to face sanitary fittings correctly.

### Measuring Pipe Lengths

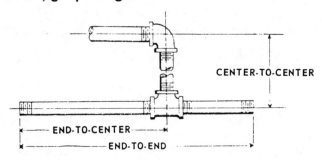

**Figure 6-15**

*Measuring pipe lengths*: Steel pipe measurements (see Figure 6–15) are always taken as follows: 1) End-to-end for a single pipe or for a run of pipe and fittings. 2) End of pipe to center of fitting. 3) Center of fitting to center of fitting. After a measurement is taken, however, an additional amount must be added on to allow for distance that the pipe will screw into a fitting. Refer to Figures 6–16 and 17 to determine this additional amount for the size of pipe you are working. Add this amount (for

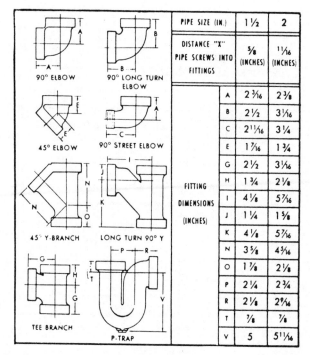

| PIPE SIZE (IN.) | | 1½ | 2 |
|---|---|---|---|
| DISTANCE "X" PIPE SCREWS INTO FITTINGS | | ⅝ (INCHES) | ¹¹⁄₁₆ (INCHES) |
| FITTING DIMENSIONS (INCHES) | A | 2 ³⁄₁₆ | 2 ⅜ |
| | B | 2½ | 3¹⁄₁₆ |
| | C | 2¹¹⁄₁₆ | 3¼ |
| | E | 1⁷⁄₁₆ | 1¾ |
| | G | 2½ | 3¹⁄₁₆ |
| | H | 1¾ | 2⅛ |
| | I | 4⅛ | 5⁷⁄₁₆ |
| | J | 1¼ | 1⅝ |
| | K | 4⅛ | 5⁷⁄₁₆ |
| | N | 3⅝ | 4⁵⁄₁₆ |
| | O | 1⅞ | 2⅛ |
| | P | 2¼ | 2¾ |
| | R | 2⅛ | 2⁹⁄₁₆ |
| | T | ⅞ | ⅞ |
| | V | 5 | 5¹¹⁄₁₆ |

**ALLOWANCES FOR THREADED DRAINAGE FITTINGS**

**Figure 6-16**

instance, ½" for ¾" pipe) to measured distance, and mark the pipe for cutting. Use a file or hacksaw to mark it.

Figure 6–18 gives you a "for-instance" measurement which will help clarify.

*Cutting steel pipe:* A pipe cutter makes the cleanest cut, but a hacksaw may be used if preferred. (See Figure 6–19.) Lock the pipe in a pipe vise, which won't crush it as a bench vise might. Thus held it can be cut and threaded. To use a pipe cutter, slip it over pipe with the wheel resting on your mark, and tighten the handle until the pipe is held firmly between wheel and rollers. Apply thread-cutting oil generously to the cutter wheel and pipe. Rotate the cutter around the pipe one turn, then again tighten the wheel against the pipe and rotate another turn. Continue in this manner ½ tightening the wheel after each rotation—until the pipe is cut through. As you near the finish, support the end being cut off to keep it from sagging or from falling when severed. Finish the cut by removing the flared metal from the pipe inside the circumference with a pipe reamer.

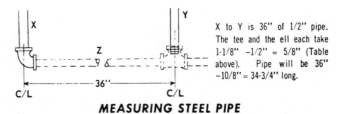

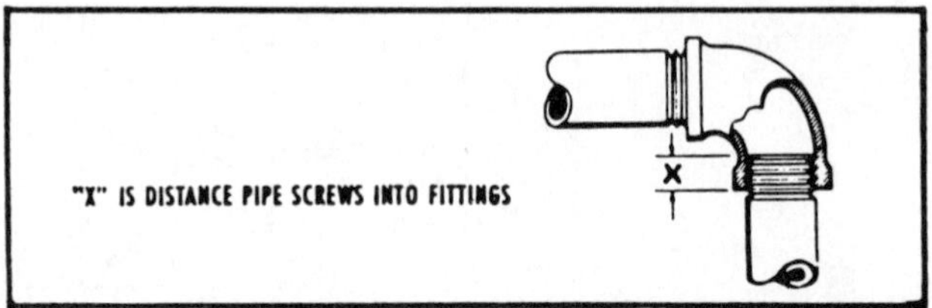
"X" IS DISTANCE PIPE SCREWS INTO FITTINGS
X

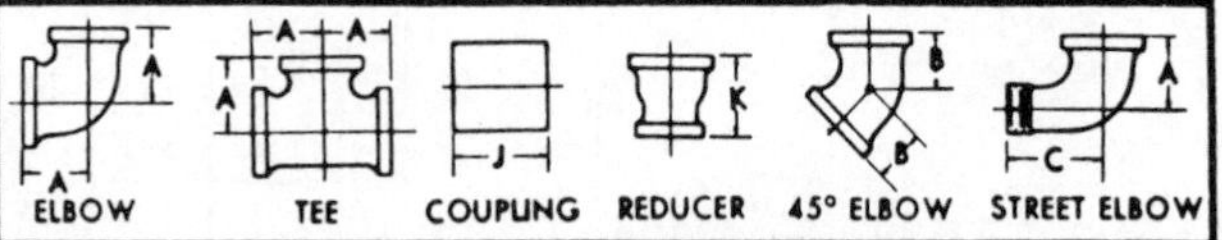
ELBOW
TEE
COUPLING
REDUCER
45° ELBOW
STREET ELBOW

| PIPE SIZE | DISTANCE "X" | FITTING DIMENSIONS (IN.) | | | | |
|---|---|---|---|---|---|---|
| | | A | B | C | J | K |
| 1/2 IN. | 1/2 IN. | 1 1/8 | 7/8 | 1 5/8 | 1 5/16 | 1 1/4 |
| 3/4 IN. | 1/2 IN. | 1 5/16 | 1 | 1 7/8 | 1 1/2 | 1 7/16 |
| 1 IN. | 9/16 IN. | 1 1/2 | 1 1/8 | 2 1/8 | 1 11/16 | 1 11/16 |
| 1 1/4 IN. | 5/8 IN. | 1 3/4 | 1 5/16 | 2 7/16 | 1 15/16 | 2 1/16 |
| 1 1/2 IN. | 5/8 IN. | 1 15/16 | 1 7/16 | 2 11/16 | 2 1/8 | 2 5/16 |
| 2 IN. | 11/16 IN. | 2 1/4 | 1 11/16 | 3 1/4 | 2 1/2 | 2 13/16 |

*ALLOWANCES FOR THREADED STANDARD FITTINGS*

**Figure 6-17**

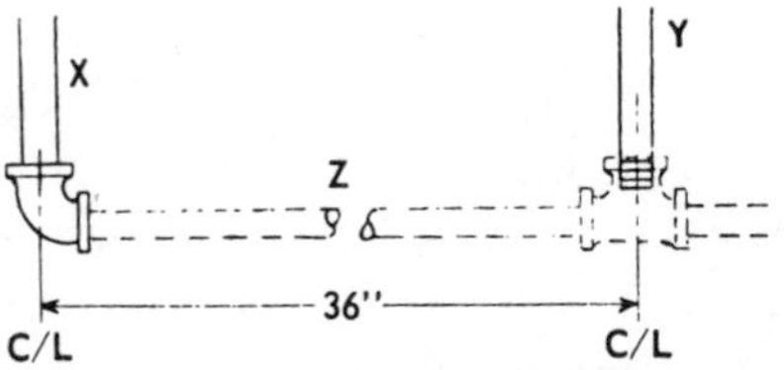
X
Y
Z
36"
C/L
C/L

X to Y is 36" of 1/2" pipe. The tee and the ell each take 1-1/8" -1/2" = 5/8" (Table above). Pipe will be 36" -10/8" = 34-3/4" long.

*MEASURING STEEL PIPE*

**Figure 6-18**

To use a hacksaw hold it squarely across the pipe on the mark, and take long uniform strokes with pressure applied only on the pull strokes. Smooth the inside circumference of the cut end with a pipe reamer (Figure 6–19), and file away any jagged edges from the outside circumference.

*Threading*: Select the die for the size pipe to be threaded and install it in the stock (handled die holder) together with a proper size guide bushing (if the stock uses a separate bushing). Since the die cuts a tapered thread the larger die opening must go over the pipe end first—and the bushing must be at this side of the die to guide it squarely into the pipe. Push the

### Cutting Steel Pipe

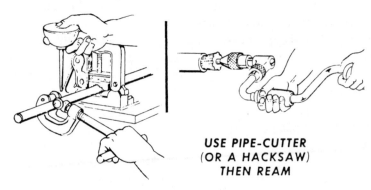

**USE PIPE-CUTTER
(OR A HACKSAW)
THEN REAM**

**Figure 6-19**

stock so that the bushing encircles the pipe and the die goes partly onto the pipe end (see Figure 6–20), then rotate the stock slowly clockwise while pushing it until the die takes hold of the pipe. Once the threads are started so that the die takes, you can stop pushing and simply continue to rotate the stock. Before continuing, however, apply cutting oil generously to the pipe and the die.

### Threading

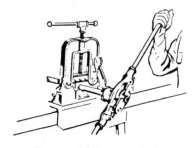

**Figure 6-20**

Rotate slowly and as continuously as possible until the pipe end projects one full thread beyond the outer die face, then back the die off by reversing the rotation. If at any time while threading or backing off, the die seems to freeze so that an excess of force is needed to rotate the stock, unfreeze it by reversing the rotation for a quarter turn or so (to clear the chips) before proceeding. When finished, clean the die for later use. *Important*: don't make too long or too short a threaded portion. The threading is finished when one full thread projects beyond the die.

*Joining pieces*: Before making any joint remove dirt and chips from the inside of the pipe and around the threads (see Figure 6–21). Apply pipe-thread compound to the pipe threads only, never to inside threads of a fitting. Use compound sparingly—just enough to fill the threads of a fitting. Use compound sparingly—just enough to fill the threads evenly with no excess or barren spots. Don't let the compound get over the end of the pipe or inside it.

### Joining Pieces

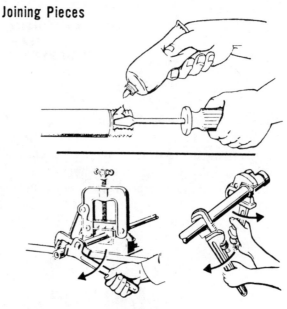

**Figure 6-21**

Start the joint and turn it closed by hand as tightly as you can, to make certain threads are tracking properly. This is especially important when starting the end of a long, heavy pipe into a fitting already installed where it is awkard to reach; under such circumstances it is quite possible to start the threads in crooked, then strip them if the wrench is used before you realize the mistake. After you're certain that threads are tracking properly, use a wrench or wrenches to draw the joint closed. Always use pipe wrenches, which are designed to tighten their grip as they are turned.

Important—a joint must be tight to be leakproof. On the other hand, a reasonably strong man can easily turn ½″ or ¾″ pipe into a fitting until the fitting cracks open. With a new fitting and new pipe threaded to depth as explained, the joint will be tight enough when about three pipe threads are still visible. Each time a joint is remade, however, the pipe will go in a little farther to achieve proper tightness. But never turn a pipe in more than one turn after the last of the thread has disappeared inside; doing so will very likely split the fitting. With a little experience you can guage the tightness by the feel.

### SWEATED COPPER PIPE OR TUBING

To measure (see Figure 6–22), measure and mark in the same manner as for steel pipe. Due to the differences in different makes of copper fittings, we do not include here a chart of fitting allowances and distances that tubing slips into these fittings. Make actual measurements of the fittings you are using to determine these allowances.

## SWEATED COPPER PIPE OR TUBING

### Measuring

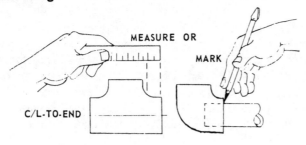

**Figure 6-22**

To cut (see Figure 6–23) sizes up to 1½", use a tube cutter; larger sizes can take a fine-tooth hacksaw blade. Do cutting the same as described for steel pipe, removing burrs with a pipe reamer or round file.

### Cutting

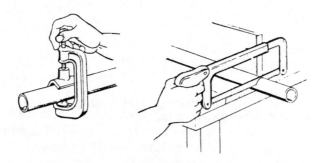

**Figure 6-23**

However, do not use cutting oil. If a hacksawed end isn't square (tubing may tear away unevenly before the cut is finished), file it square with a flat file.

*Note*: The best way to hold pipe or tubing for cutting is to lay it between two boards clamped to a bench, as in Figure 6–23. A vise (even a pipe vise) is likely to crush it if it is tightened too much. If a vise is used, let it grip the pipe at least 6 inches from the end so that any crushing by the vise will not cause the end that fits into the fitting to become out-of-round. If the end is out-of-rounded you probably can't make a successful joint—it would be best to cut it off to where the pipe is round.

*Joining*: A tubing or pipe end and a fitting are joined by slipping the end all the way into the fitting hub—until it bottoms against the hub shoulder—then filling the tiny space between the sides of the hub and tube with a thin layer of solder to "weld" the parts together. It is very easy to make such a joint if you obey these basic rules:

1. Both the inside of the fitting hub and the end of the tube must be perfectly round, unmarred by nicks or scratches, and brightly clean. A nick or deep scratch may leave a space too big for the solder film to fill. Any dirt, tarnish or oil (even the oil left by a fingerprint) will prevent the solder from becoming "welded" to the copper. Brighten both surfaces carefully with emery cloth (Figure 6–24) or fine steelwoll and wipe with a clean, lintless cloth—then don't finger them.

**Joining**

**Figure 6-24**

2. Surfaces must be coated with correct type of flux. Fluxing accomplishes several purposes. It finishes the cleaning job to prepare the surfaces for binding to the solder—and it wets the surfaces so that the solder will flow evenly over them. Use only a non-corrosive (not an acid type) flux, either liquid or paste, and coat both surfaces thinly and evenly. Then slide the tube end into

the fitting and twist it a turn or so to make certain the flux is well distributed. Leave the tube in the fitting. (See Figure 6–25.)

COAT SURFACE
WITH FLUX

**Figure 6-25**

3. Proper type solder must be used. Use only solid-core (not flux-core) wire solder of 50-50 (half lead, half tin) type.

4. The tube and fitting must be heated to the temperature at which they will just melt the solder wire and suck the solder into the joint. If the flame is applied to the wire solder it will quickly melt—even though the tube and fitting are still cold—but it will not flow into the joint and would not "weld" to the surfaces even if it did. (See Figure 6–26.) Also, if the tube and fitting are overheated they will melt the wire as fast as you feed it to them and the excess will drip out so that you will have difficulty knowing whether or not a sufficient amount remains in the joint to thoroughly seal it.

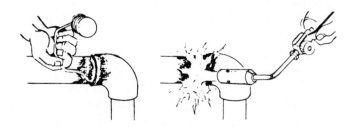

**Figure 6-26**

The heating capacity of the flame required to heat a joint for soldering depends upon how fast heat is conducted away from the joint. The tube size, length and starting temperature are all factors. For instance, it may take twice as much flame and a longer time to heat a joint in a long run that extends outdoors into winter cold as to heat the same joint if the tube is short (not part of a run) and is entirely within a warm basement. Usually, a propane torch is satisfactory for heating ½" or ¾" tube joints;

but larger size tubes and pipes require the larger-capacity flame of a gasoline blowtorch.

For ½″ or ¾″ joints apply the flame to the tube and fitting (moving it from one to the other) on just one side (if the joint overhead, the bottom side). As heat accumulates the spots licked by the flame will start to change color and the flux will bubble and spit from the joint. When they do, move the flame onto the fitting and touch the end of the solder wire to the joint crack at the side opposite the flame, being careful not to allow the flame to lick the wire. If the solder does not instantly melt into the crack, continue to hold the flame on the fitting until it does. At the instant when the solder does flow readily into the crack, remove the flame—and feed solder until a small bead of it gathers at bottom side of the crack, then remove it. If the bead hangs on, you have completed a perfectly made joint. Should it drip off, the joint is too hot and solder is running out. You will have to feed more solder, until the forming of a permanent bead assures you that the joint remains filled. If no bead appears, the joint has cooled too quickly and needs to be heated again—until the bead does form. Always let the solder cool and set before jarring or moving the joint.

For larger size joints the above also applies, with this exception: if the flame is applied only to one side, this side may be overheated while the opposite side will be insufficiently heated. *Best thing to do*: move the flame around both sides of the joint as far as possible. Otherwise, hold the flame until the opposite side does become sufficiently heated, then remove the flame and wait a second or two for the heat to become evenly distributed around the joint before feeding in enough solder to fill the joint.

*Caution*: If wood (a joist, etc.) or other combustible material is within reach of flame when soldering, shield it with a piece of asbestos board or sheetmetal. Wood that may appear to you to be merely charred can often later be fanned into a fire. If a blowtorch is used, remember that gasoline can be dangerous if improperly handled. Keep the container tightly closed when not refilling the torch and, preferably, in another part of the building (or out in the garage) from where you are using the torch. Don't light the torch in the presence of the gasoline container, and be sure that your torch gasoline cap is securely on.

Much time can be saved by preassembling an entire run (with cleaned, fluxed joints all fitted together) as far as possible, then soldering all joints in a continuous operation. This way, for instance, the two joints at an ell usually can be made with one heating.

If you are a novice and feel uncertain about soldering joints as described, a slower process called "tinning" will assure you of success. This is the precoating of both surfaces (pipe and fitting) with a thin layer of solder. To tin, clean and flux each surface. Take the pipe into a vise, and apply a flame about 2″ from the end until you can melt a drop of solder onto the top and side of the end. Quickly remove the flame and solder wire, and using a lintless rag pad (thick enough not to burn your hand) wipe the molten drop to spread it thinly all over the pipe end that will be inside the fitting. If necessary, rotate the pipe 180°, reheat it, and try again until the solder is properly spread. To tin the fitting, hold it in a vise at a 45° angle, heat it, apply a drop of solder to the bottom inside of the hub, then twist a finger of rag around the inside to spread the solder thinly around the entire hub surface. Don't allow the solder to run down the inside of the fitting or to form too thickly at any point.

*Note*: You can't very well tin one side of an already installed fitting. Before installing it, tin all its hubs.

With solder prebonded to both surfaces by tinning, all that remains to be accomplished when assembling the joint is to fill it with solder. As the tinning probably will interfere with slipping pipe into a fitting, heat both parts until tinning softens and you can bottom the pipe in the fitting—then quickly apply enough solder to fill the joint. You can be sure the joint will be properly done if the tinning is thus softened. Also, it will help to preassemble as many joints as possible on the floor or the workbench—then lift the entire assembly into place for joining to its run.

### FLARE-JOINT TUBING

*Note*: The flexible tubing with which this type of joint may be used is measured and cut as described preceding for sweated joint copper pipe or tubing.

Flaring is done with a flaring tool. First, however, remove the nut from fitting and slip it, threaded side facing out, over the tubing end (it can't be put on after the tubing is flared). Hold the tubing end (see Figure 6–27) and flaring tool as shown, and tap the tool with a hammer until the tubing end is spread out enough so that the threaded side of the nut will just slip out over it. Care must be taken to spread it uniformly all around.

*Joining*: Be sure there are no burrs or dirt to interfere with smooth contact, then hold the flared tubing end squarely against the fitting end (see Figure 6–28), and start the nut onto the fitting threads by hand. Tighten the nut as much as possible by hand to make certain threads are tracking correctly. Finish by tightening it securely, using two wrenches:

### Flaring

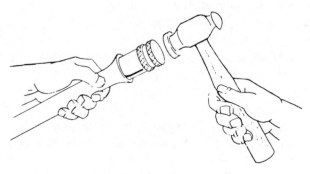

**Figure 6-27**

### Joining

**Figure 6-28**

one on the nut, the other to hold the fitting. Never tighten with one wrench alone; if the fitting isn't held, it may turn and spoil the connection at its other side.

## PLASTIC DRAINAGE PIPE

Measuring and cutting plastic pipe is the same as for sweated copper pipe, except that a fine-tooth hacksaw is used for all pipe sizes. To join plastic pipe (see Figure 6–29), wipe the pipe end and the inside of the

### Joining

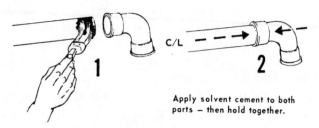

Apply solvent cement to both parts — then hold together.

**Figure 6-29**

fitting hub clean with a dry, lintless rag. Surfaces must be clean and dry. Use a small paint brush and the special solvent cement made for this purpose. Paint the entire contact surface inside the fitting hub with a thin coating of the cement. Also, paint the entire contact surface of the pipe end and slide the pipe end into the fitting hub—at once, before the cement dries—and then hold it steady until the cement around the crack dries, in a minute. Do not jar, turn or wet the joint for at least another hour, in which time the cement will harden to permanently weld the joint. *Note*: wipe the brush clean and then the cement can when it's not in use.

### FIXTURE DRAINS

Note that fixtures must be installed first.

*Measuring and cutting*: A lavatory drain or sink tailpiece should extend down into the hub at the trap top as far as possible, but a good joint can be made with as little as ¼" of it extending into the hub. Drop the drain (or loosely fasten the tailpiece) in place and hold the trap up at the level at which it must be to mark the cut. At the same time you can also mark the other trap end for cutting off, if necessary. This end should slide into a nipple joined to the branch drain so as to project ½" to 1" from the wall (or floor) and it should slide in at least ¼" to 1", but no more than the length of the nipple.

The chrome-plated stainless steel or brass tubing used for fixture drains is very thin walled and can be easily bent or crushed. The part around which the washer and nut will fit must be round. Therefore care must be taken to hold it carefully while cutting—not in a vise, preferably by hand, or in a channel between two boards as for copper pipe. Cut it with a fine-tooth hacksaw, without using enough force to distort it. Afterwards, rub the cut edge with emery cloth to remove any burrs. (See Figure 6–30)

*Joining*: With the slip nut disengated from its threads, slide the plain tube end through the nut (from its back side) and install the rubber washer on the tube end. Insert this plain end into the hub (or nipple) as far as it is to go, adjusting the washer to just contact the edge of the hub or nipple. See Figure 6–31. Make certain the washer isn't twisted. Start and tighten the nut by hand, then finish with a tool. To prevent marring it, don't use a pipe wrench; use a monkey wrench or correct-size open-end wrench. If tightening trap connection, hold the trap steady with your other hand. Make the joint fairly tight (the washer must be squeezed into the crack), but not so tight as to split the nut or to strip the threads. Note that if an escutcheon is used it must be placed on the tube ahead of the nut.

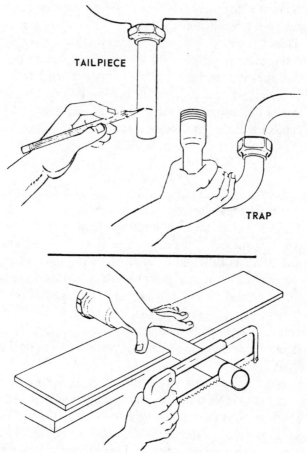

TAILPIECE

TRAP

Figure 6-30

## Joining

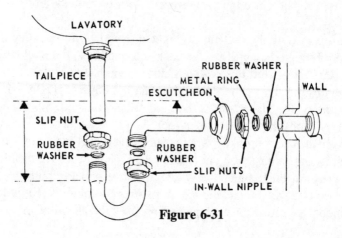

LAVATORY

TAILPIECE

RUBBER WASHER

METAL RING

ESCUTCHEON

WALL

SLIP NUT

RUBBER WASHER

RUBBER WASHER

SLIP NUTS

IN-WALL NIPPLE

Figure 6-31

## FIXTURE SUPPLY LINES

Note that the fixture must be installed first. These lines usually are not measured and cut, as they are purchased in correct lengths. Also, they are bendable, flexible, to allow considerable variation.

*Slip-nut fittings*: Some fixture supply lines are furnished with the ends flared and slip nuts in place, ready for attachment as explained for copper tubing flare-type joints. See Figure 6–32.

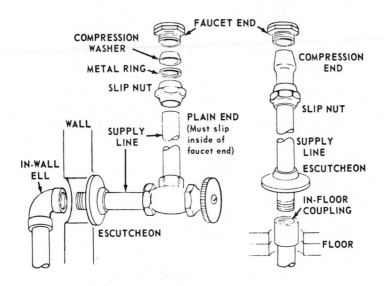

### Compression-Nut Fittings

**Figure 6-32**

*Compression-nut fittings*: Lines so fitted are connected in the same manner as the fixture drains, preceding—except that, instead of rubber washers, they have soft brass washers. Tube must extend at least 1″ into the water-branch nipple. If the joint has to be remade, a new washer is required.

### PLASTIC WATER PIPE (FOR OUTDOORS)

One type of plastic cold-water pressure pipe—used with wells, sprinkler systems, etc., and approved for drinking water—is joined with fittings similar to those used for plastic drainage pipe, with a special PVC cement.

The high-pressure pipe (used for cold water runs other than those inside a home and for wells) uses special high-impact fittings which are clamped like hose fittings into place. See Figure 6–33. Pipe, furnished in cut lengths and in coils, can be cut off with a sharp knife or a fine-tooth hacksaw. It is pushed over the end of a fitting far enough to be gripped by two stainless-steel pipe clamps. (Clamps should be in position, first.) If you have difficulty twisting it on over the fitting, hold the pipe firmly and tap the fitting in with a rubber mallet (don't use a hammer or you will shatter the fitting).

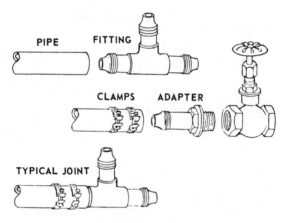

**Figure 6-33**

### FIBER PIPE (FOR OUTDOORS)

*Measuring and cutting*: Use the same measuring methods as for cast-iron pipe, preceding—however, close measurements are seldom required. To cut, use a coarse-tooth crosscut hand saw. Cut squarely across. (See Figure 6–34.)

*Joining*: The unperforated fiber pipe requires tight joints, and is furnished in lengths having tapered ends to slide into any standard fittings. Simply place the fitting (a coupling, ell, tee, etc.,) over the tapered end and drive it tightly on with a hammer, using a wood block as a buffer. If pipe is cut, the taper is lost at this end. In such case it must be joined to a tapered-end pipe length by using an adapter (which has one tapered opening and one untapered one) or to another cut. End the pipe length by using a joining sleeve which has two untapered sides. See Figure 6–35. Since neither the adapter nor the joining sleeve can be used to join pipe to one of the standard fittings, it is necessary to plan your run so that only tapered pipe ends will be where a fitting such as an ell, tee, etc., is needed.

## FIBER PIPE (for Outdoors)
## Measuring and Cutting

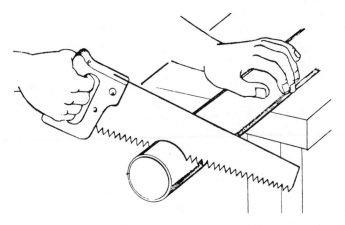

**Figure 6-34**

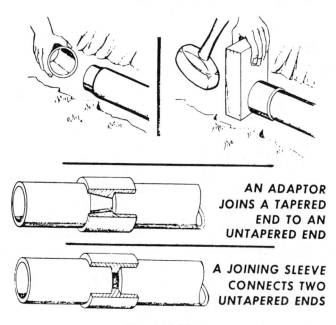

AN ADAPTOR
JOINS A TAPERED
END TO AN
UNTAPERED END

A JOINING SLEEVE
CONNECTS TWO
UNTAPERED ENDS

**Figure 6-35**

The above fiber pipe can be joined to the spigot end of cast-iron pipe by use of the adapter or the joining sleeve, depending upon whether fiber pipe end is tapered or a cut end. Either a tapered or a cut end can be fitted directly into the hub of a cast-iron pipe. A joint of this type is sealed by packing it partially with oakum, then caulking it with lead wood (refer to cast-iron pipe joining) or filling it with a plastic sealing compound, asphalt or cement. Do not use molten lead.

Perforated fiber pipe doesn't require water tight joints (see Figure 6–36); all joints are made with simple snap couplings.

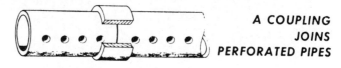

**A COUPLING
JOINS
PERFORATED PIPES**

**Figure 6-36**

## VITRIFIED TILE PIPE (FOR OUTDOORS)

*Measuring and cutting*: Do these both in the same manner as for cast-iron pipe. However, this pipe is exceedingly brittle. For cutting, use a sharp cold chisel, and tap very lightly around and around until the pipe

**Joining**

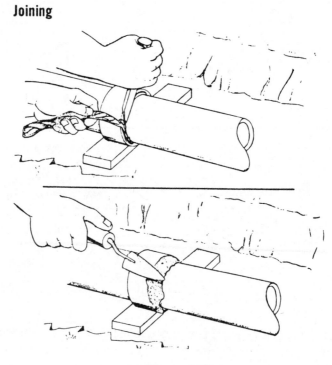

**Figure 6-37**

is cut off. If possible, have the area being cut supported in a bed of sand, to protect it against stress. See Figure 6–37.

*Joining*: When used for house gutter drainage the tile pipes can be assembled by simply laying them together, spigot in hub, without sealing. In fact, so doing is a good way of spreading rain water through a lawn. However, such joints will permit tree and shrubbery roots to grow into and eventually clog the pipe run, and some plants will send root out 20′ or more if they can "smell" a source of water. If you anticipate such a problem, seal the joints as follows: pack it half full of oakum, then fill it with cement. Use a 1 to 2 (one part cement, two parts sand) mix, with just enough water to create a putty-like paste. To make it easier to work, prop the hub on a 2 × 4 until the joint is finished, then slide the 2 × 4 out to let the joint settle in place, and reach around the joint to make certain that cement is still packed in place.

# 7

# PLANNING NEW INSTALLATIONS

One of the major advantages of doing a high percentage of your own work on a new installation is that you can achieve a higher level of luxury at a lower cost than you can through the use of outside sources. For example, you can install an island kitchen sink, outdoor faucets, a garden fountain and pool, additional bathrooms, garage faucets, a basement shower, an extra enclosed shower stall, a lawn sprinkler system, hobby area faucets, a dehumidifier, and other conveniences not ordinarily seen in household plumbing installations.

A major source of help is your local supplier. Wards, for example, offers aid to anyone who is designing or who plans to contract or subcontract plumbing installations. It isn't necessary to prepare detailed piping plans or accurately count pipe lengths and fittings. Wards suggests ordering generously, and offers to take back for full credit all unused parts.

A water-supply system does not require any special planning. Pipes can easily be cut and fitted in wherever needed, on the job. Aside from your personal preferences for certain models, colors and quality of trim, the only other requirement is to make certain each fixture will fit into the space alloted for it. Check the overall dimensions of each fixture if there is any doubt.

Some things to keep in mind when planning your system which will save you money and make for greater efficiency:

1. Locate new fixtures as close to each other as possible, so less pipe and fewer fittings will be needed.
2. The most expensive part of a new plumbing system in a home is

the soil stack, so try to locate your fixtures so that a single stack will do the job.

3. If you are using a septic tank, install three lines instead of the customary single line. Use your septic tank for only solid or heavy waste material. Run the second line from the washing machine to a seepage pit to prevent septic tank overloading. The third line could run from the kitchen sink to a grease trap and from there to the septic tank line. This system will prevent overloading, and cut down on cleanout costs.

4. Plan for back-to-back installations wherever possible so that the same lines, for example, might serve both the kitchen and the bath.

5. Think of the future by adding a few tees when running a long line, plugged for the present, but a future money-saver when you may want to tie in new lines.

6. Locate your water heater as close as you can to the hot water faucets it will serve so there will be less heat loss.

7. Be certain you have no leaks in the drain system from which sewer gas might enter the home.

8. Be sure you have no hook-ups between the water supply system and the drain system which could bring unwanted waste into a fixture outlet.

9. Be certain fixtures are installed so as to prevent any back siphonage from the fixture into the water supply system.

10. You can save bath floor area by using stall showers instead of tubs.

11. Locate bath entrances so as to be unseen from the living or dining room.

12. Don't forget to plan for adequate heating, lighting and ventilation in bath and kitchen areas.

13. Plan so a suddenly opened bathroom door will not strike anyone using a fixture.

14. In a two-story home, fixtures should be located in a continuous vertical line, and arrange all fixtures in a straight line, for efficiency and savings.

## CHECK YOUR CODE

If your community has a plumbing code you probably can obtain a copy at your city hall or county courthouse. Codes relate principally to the installations of drainage systems, and typical areas of code application are as follows:

1. *Kind of drainage pipe permitted for unburied use.* Some codes require a certain kind (cast-iron and steel or copper, etc.); other codes permit any one of several kinds. Still others specify one material for main stacks, but allow a choice for waste and vent lines, or for secondary stacks. (See Figure 7–1 for typical pipe plan code-related materials.)

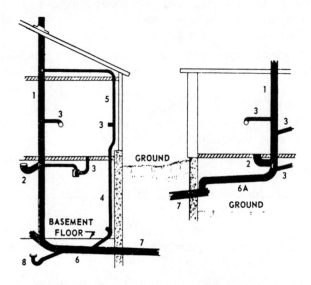

1 – MAIN STACK – Cast Iron, Plastic, Copper – 3" or 4".
2 – TOILET BRANCH – Same as Main Stack.
3 – LAV., TUB, SINK BRANCHES – Steel, Plastic or Cop-
      per – 1½".
4 – SECONDARY STACK – Steel or Copper – 1½".
5 – REVENT RUN – Steel, Plastic, Copper – 1½".
6 – HOUSE DRAIN – Cast Iron – Size of Main Stack or Larger.
6A – SUSPENDED HOUSE DRAIN – Cast Iron, Plastic or Cop-
      per – Size of Main Stack or Larger.
7 – HOUSE SEWER – Cast Iron . . . Sometimes Plastic, Fiber,
      Tile or Concrete – Size of House Drain.
8 – FLOOR OR SIMILAR DRAIN – Cast Iron – 2".

**Figure 7-1**

2. *Kind of drainage pipe that is permitted for use where pipe will be buried underground or in concrete, etc.* Some codes specify one kind of pipe for use under a building, but permit use of several different kinds for that part of the system, such as the house sewer, which is outside of the building. Generally, only cast-iron is allowed within the building area (the house drain), and is preferred for outside (house sewer) if there is any danger of crushing. Ordinary copper drainage pipe is never buried; plastic,

fiber, tile and other types are permitted outside (house sewer) under proper conditions.

3. *Venting requirements.* Every code requires a drainage system to be vented by extension of the main stack through the roof. Generally, in cold climates where frost might clog a small pipe, the

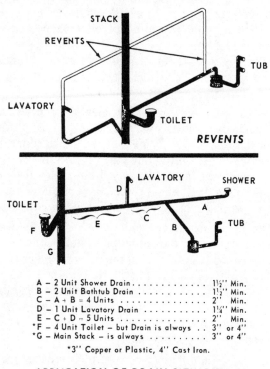

**REVENTS**

A – 2 Unit Shower Drain . . . . . . . . . . . . . 1½'' Min.
B – 2 Unit Bathtub Drain . . . . . . . . . . . . 1½'' Min.
C – A + B = 4 Units . . . . . . . . . . . . . . . 2'' Min.
D – 1 Unit Lavatory Drain . . . . . . . . . . 1¼'' Min.
E – C + D = 5 Units . . . . . . . . . . . . . . . 2'' Min.
*F – 4 Unit Toilet – but Drain is always . . 3'' or 4''
*G – Main Stack – is always . . . . . . . . . . 3'' or 4''

*3'' Copper or Plastic, 4'' Cast Iron.

**APPLICATION OF DRAIN SIZING RULES**

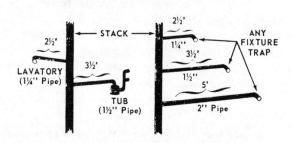

**MAX. HORIZONTAL DRAIN LENGTHS FOR
VARIOUS PIPE SIZES, WITHOUT REVENTING**

**Figure 7-2**

above-roof portion of any vent is required to be up to 4″ diameter pipe, and there may be further specifications regarding the height above the roof, etc. Some codes require no more venting than this, while others require certain fixtures—depending upon where and/or from what distance they are connected to the stack—to be separately revented either out through the roof or by connection to the stack at an elevation above all waste connections. Still other codes require reventing of all lavatories, tubs and sinks without exception.

(See Figure 7–2 for diagrams showing revents, application of drain size rules, and maximum horizontal drain lengths for various pipe sizes without reventing.)

4. *Fixture waste connection points*. Generally, each fixture is separately connected by its own waste line to the vertical stack. In some instances, however, it is permissible to connect two fixtures into one waste line, providing the common line is large enough to carry both wastes simultaneously. Most codes also allow a fixture (or some types of fixtures) to be connected to a horizontal house drain, if venting requirements are properly complied with.

5. *Trap requirements*. Every code requires that the waste line connecting a fixture to the drainage system contain a water trap. Toilets have built-in traps, but the traps for other fixtures must be separately provided. A trap is so designed that some of the waste-water flow will always remain to keep the trap full (unless, during a period of disuse, the water evaporates). Thus, there is always a plug of water in the line to prevent system gases from rising through the fixture outlet.

6. *Kind of water-supply piping*. As a rule, threaded galvanized steel pipe and rigid or flexible copper tubing with soldered joints are specified for in-house use, but not for use underground or in concrete. For the latter purpose a special heavy-duty copper tubing (with soldered or flare-type joints) is preferred, though lead pipe also may be specified. Some codes permit use of plastic pipe for cold-water only runs, under certain conditions. Also, steel pipe may be used underground outside of the building area where it can easily be dug up, if necessary.

In the absence of any local code to the contrary we recommend adopting the provisions of the National code, which is summarized in Figure 7–3.

Figure 7–4 shows correct and incorrect methods of fixture placement and installation, and offers installation information and tips.

## SUMMARY OF NATIONAL CODE PROVISIONS RECOMMENDED
## FOR AVERAGE HOME APPLICATION

If there is a bathroom there must be at least one main stack.

Any main stack should be at least 3-in. dia. pipe (if hub-and-spigot cast-iron pipe is used, preferably 4-in. dia.) throughout its entire length, including the vent portion that extends through the roof.

The waste portion of any secondary stack should be at least as large in dia. as the largest branch drain it serves. The vent portion should be at least 1-1/2-in. dia., and should either be vented (extended through the roof) or be revented (connected to a main stack).

That portion of any house drain that serves a particular stack should be at least as large in dia. as the stack. If any portion serves two or more stacks, this portion should be at least 3-in. dia. (preferably 4-in., if hub-and-spigot cast-iron pipe is used).

A house sewer should be at least 3-in. dia., preferably 4-in.

The trap and individual drain for any fixture is sized according to the fixture rating.

| FIXTURE | UNIT RATING | TRAP | DRAIN |
|---|---|---|---|
| Toilet with flush tank ...... | 4 | Built | *4" or †13" |
| Toilet with flush valve ...... | 6 | In | |
| Bathtub ................ | 2 | — 1-1/2" — | |
| Shower.................. | 2 | — 1-1/2" — | |
| Lavatory ............... | 1 | — 1-1/4" — | |
| Sink ................... | 2 | — 1-1/2" — | |
| Sink with Disposer......... | 3 | — *2" or †1-1/2" — | |
| Laundry Tub ............. | 2 | — 1-1/2" — | |
| Floor Drain ............. | 1 | — 1-1/4" — | |

*Size if using hub-and-spigot cast-iron.
†Size if using any other kind of pipe.

**Figure 7-3**

Figure 7–5 shows typical drainage systems available from Wards and other sources of plumbing supplies.

### THE FIRST STEP—LOCATE THE STACK(S) AND FIXTURES

The costliest single item in a plumbing system is a main soil stack. And a main stack also can create construction problems due to its large diameter. A 4" hub-and-spigot cast-iron stack, for instance, requires a 6¼" space. That is, the studs of the partition in which it is located must be 2 × 8s or two 2 × 4s edge-to-edge, whereas partitions generally are built with 2 × 4 studs or, in some modern homes, even 2 × 2 studs. Therefore, a stack can necessitate additional construction cost, and can waste space.

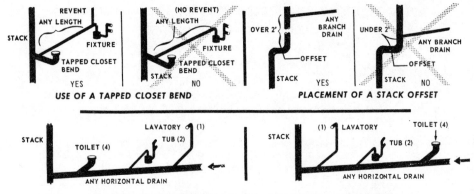

**USE OF A TAPPED CLOSET BEND**                    **PLACEMENT OF A STACK OFFSET**

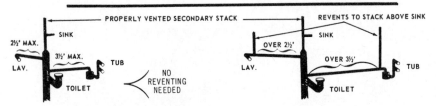

**CORRECT — SEQUENCE OF FIXTURE CONNECTIONS — INCORRECT**

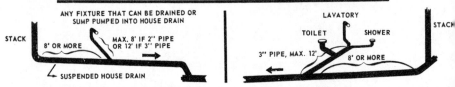

**CONNECTING BASEMENT FIXTURES TO FOOT OF A SECONDARY STACK**

**TYPICAL UNREVENTED BASEMENT FIXTURE INSTALLATIONS**

To figure the size of a horizontal branch drain that serves two or more fixtures add the fixture ratings and apply the total as follows. Note, however, that the total for a bathroom group (toilet, lavatory and tub or shower) is considered to be 6.

| RATINGS TOTAL | PIPE SIZE |
|---|---|
| 1 | 1-1/4" |
| 2 to 3 | 1-1/2" |
| 4 to 6 | 2" |
| 7 to 12 | 2-1/2" |
| *12 to 20 | 3" |

*Not to include more than two toilets, however.

The rule for reventing of fixtures connected by horizontal branch drains to a stack is dependent upon the size of the branch drain. Table below gives the maximum length of pipe between fixture and stack . . . if pipe exceeds this length the fixture should be revented.

| PIPE SIZE | MAX. LGTH. |
|---|---|
| 1-1/4" | 2-1/2 ft. |
| 1-1/2" | 3-1/2 ft. |
| 2" | 5 ft. |
| 3" | 6 ft. |
| 4" | 10 ft. |

A tapped closet bend can be used for connection of the branch drain from another fixture if the other fixture is revented . . . but cannot be used for connection of a vent line. Since a toilet is seldom more than 24 in. from its stack, reventing of

it is not customarily required. Should it be necessary, however, then the toilet branch drain must be extended "upstream" from the toilet to provide a connection for the revent line.

Whenever several fixtures are connected to the same horizontal branch drain (or house drain) each smaller rated fixture should be connected ahead of the larger rated ones. Hence, if one of the fixtures is a toilet (largest rating) its connection should be the last one on the "downstream" (direction of flow) end.

No branch drain should join a stack within 2 ft. of any offset in the stack.

Fixtures (including a toilet) located in a basement may be connected to a secondary stack that is increased in size from point of connection downward to a sufficient size to accommodate them . . . and do not require reventing if above pipe max. lengths are complied with.
A toilet and/or other fixtures located in a basement also may be connected to a horizontal house drain of 3-in. or larger dia. without reventing, if: 1) The connection is 8 ft. or more distant from the foot of the nearest stack served by the house drain . . . and 2) The branch drain connecting the fixture(s) to the house drain is not over 8 ft. long if 2-in. pipe or 12 ft. long if 3-in. pipe. Otherwise, provide a revent (or revents).

If it is permitted to connect a storm (areaway or gutter) drain to the sanitary sewer system, the storm drain should join your house sewer 10 ft. or more distant from the last other connection. A gutter drain so connected should be vented unless the drain is air tight all the way up to the roof.

**Figure 7-4**

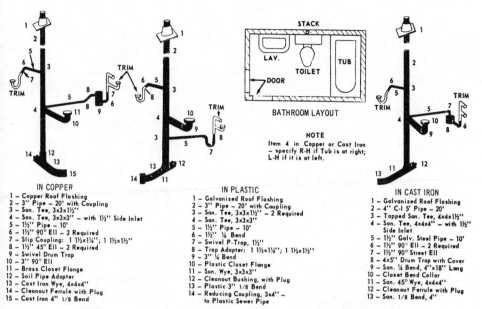

**BATHROOM LAYOUT**

**NOTE**
Item 4 in Copper or Cast Iron
– specify R-H if Tub is at right;
L-H if it is at left.

**IN COPPER**
1 – Copper Roof Flashing
2 – 3" Pipe – 20' with Coupling
3 – San. Tee, 3x3x1½"
4 – San. Tee, 3x3x3" – with 1½" Side Inlet
5 – 1½" Pipe – 10'
6 – 1½" 90° Ell – 2 Required
7 – Slip Coupling: 1 1½x1¼"; 1 1½x1½"
8 – 1½" 45° Ell – 2 Required
9 – Swivel Drum Trap
10 – 3" 90° Ell
11 – Brass Closet Flange
12 – Soil Pipe Adapter
13 – Cast Iron Wye, 4x4x4"
14 – Cleanout Ferrule with Plug
15 – Cast Iron 4" 1/8 Bend

**IN PLASTIC**
1 – Galvanized Roof Flashing
2 – 3" Pipe – 20' with Coupling
3 – San. Tee, 3x3x1½" – 2 Required
4 – San. Tee, 3x3x3"
5 – 1½" Pipe – 10'
6 – 1½" ¼ Bend
7 – Swivel P-Trap, 1½"
8 – Trap Adapter: 1 1½x1¼"; 1 1½x1½"
9 – 3" ¼ Bend
10 – Plastic Closet Flange
11 – San. Wye, 3x3x3"
12 – Cleanout Bushing, with Plug
13 – Plastic 3" 1/8 Bend
14 – Reducing Coupling, 3x4" –
    to Plastic Sewer Pipe

**IN CAST IRON**
1 – Galvanized Roof Flashing
2 – 4" C-I 5' Pipe – 20'
3 – Tapped San. Tee, 4x4x1½"
4 – San. Tee, 4x4x4" – with 1½"
    Side Inlet
5 – 1½" Galv. Steel Pipe – 10'
6 – 1½" 90° Ell – 2 Required
7 – 1½" 90° Street Ell
8 – 4x5" Drum Trap with Cover
9 – San. ¼ Bend, 4"x18" Long
10 – Closet Bend Collar
11 – San. 45° Wye, 4x4x4"
12 – Cleanout Ferrule with Plug
13 – San. 1/8 Bend, 4"

**BASIC PACKAGES – FOR 1ST-FLOOR BATHROOM AND STACK, UNREVENTED**
(Order Basic Pkg. in Copper, Plastic or Cast-Iron to receive all listed parts. Order other parts as follows.)

**WITH COPPER** – Cast-Iron Wye Assy. is furnished with Basic Pkg. –
simply order needed 5' lengths of 4" Cast-Iron Pipe.

**WITH PLASTIC** – Return parts 12 to 14. Order 3x4" Reducing Bushing
and Plastic to Hub-End C-I Pipe Adapter for use at A – also order parts
11, 12 and 13 from C-I Basic Pkg. plus needed 4"x5' C-I Pipe.

**WITH CAST IRON** – Same as with Copper, above.

**WITH COPPER** – Move Cleanout Wye Assy. to in-wall position – and
order 2 3" 90° Copper Ells (A) plus needed 3"x10' Copper Pipe.

**WITH PLASTIC** – Order needed 4"x10' Plastic Sewer Pipe to run from
Cleanout Wye Assy. through wall as in illus. at left. If House Sewer
will be C-I order 4" Plastic to Hub-End C-I Adapter to use at C.

**WITH CAST IRON** – Like Copper above except need 2 San. 4" ¼ Bends
(A) and 4"x5' pieces of C-I Pipe.

**ADDING UNDERGROUND (At Left) OR SUSPENDED (At Right) HOUSE DRAIN**

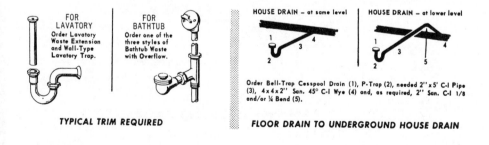

**FOR LAVATORY**
Order Lavatory
Waste Extension
and Wall-Type
Lavatory Trap.

**FOR BATHTUB**
Order one of the
three styles of
Bathtub Waste
with Overflow.

Order Bell-Trap Cesspool Drain (1), P-Trap (2), needed 2"x5' C-I Pipe
(3), 4x4x2" San. 45° C-I Wye (4) and, as required, 2" San. C-I 1/8
and/or ¼ Bend (5).

**TYPICAL TRIM REQUIRED**

**FLOOR DRAIN TO UNDERGROUND HOUSE DRAIN**

**Figure 7-5**

A secondary stack—in addition to a main stack—does, of course, add to the drainage system cost. But, since a smaller pipe size is used, a secondary stack costs less than a main stack—and creates no major construction problem. Even in hub-and-spigot cast-iron pipe a 2″ stack can be run up through a 2 × 4 partition if a ½″ furring strip is nailed to the studs. Also, the principal purpose of a secondary stack is to serve a kitchen sink, which usually is backed up against an outside wall. In such case the secondary stack is fitted into a chase (indentation) provided as the wall is erected.

It follows that the most economical (for plumbing) house plan is one requiring only a main stack, and next in economy is one that requires just one main stack and one secondary stack.

For a one-story house to have just a main stack the bathroom and kitchen must be back-to-back or side-by-side, and the kitchen sink probably will have to be against a partition instead of an outer wall, unless the outer wall location is only 5 or 6′ from the partition containing the stack. If such a ground-floor plan can be arranged, a second bath (on a second floor or in the basement) or a laundry tub in the basement also can be served by this single stack, if located close enough to the stack above or below the ground-floor bathroom.

If one main and one secondary stack are used, the kitchen can be located wherever desired, even at the opposite end of the house from the bathroom. With a plan like this in a one-story house the one main stack can be made to serve two back-to-back bathrooms and/or another bathroom or laundry tub in the basement below. Or, the laundry tub, a shower and/or lavatory in the basement could be served by the secondary stack. In a two-story house these two stacks will permit any of the foregoing arrangments plus one or two back-to-back bathrooms on the second floor above the ground-floor bathroom. In such cases the ground-floor bathroom could probably be a half bath, or powder room.

Wherever there is a basement with the house drain under the basement floor, it generally is possible to plan for any type of plumbing fixtures desired in the basement, to empty into either the stack or into the house drain, without the additional cost of reventing these fixtures. Even where there is a suspended house drain (one above the elevation of the basement floor), the same may be possible if a waste pump is used to lift the waste up to the house drain.

Additional main stacks are needed only when there are two or more bathrooms (or half baths) separated horizontally by distances which make it impossible to serve them with one stack. Additional secondary stacks are needed only when fixtures other than toilets are similary separated from each other and the bathroom. The practical maximum distance between a toilet drain opening and its main stack is only about 30″ (usually limited to 16″) except, as in a basement, where there may be room to pitch the waste run as required. For all fixtures the practical maximum distance depends upon the branch drain pipe size and the reventing requirements of your local code.

## FIXTURES NEEDED FOR CONVENIENCE

There is, of course, no absolute rule as to how many bathrooms and/or powder rooms (a toilet and lavatory) a home should have. Most builders now consider the minimum to be a full bathroom for each two bedrooms. Also, if you do much entertaining, a powder room separate from the family quarters is a great convenience. Whether you install a tub, a shower or a combination in each bathroom is a matter of personal preference.

Of equal importance with bathrooms are the kitchen facilities. The minimum requirement is, of course, a sink. In a large kitchen there may also be a second and smaller sink, for vegetable preparation or the like. Modern kitchens generally include a garbage disposer—and you may also want a dishwasher or an automatic ice-cube maker, which requires only a cold-water line. And every house should have some provision for a laundry, whether this will contain only a tub or the latest automatic equipment. Also, don't overload the convenience of outdoor cold-water hose connections for lawn and garden watering and other uses. And, should you be planning a swimming pool, think of how easy it will be to fill it through a ¾″ cold-water pipe direct to the filter, and to drain it (if your code permits) into your house drain.

Other appliances for you to consider are:

1. A hot-water heater of ample capacity (a minimum of 30 gals. size plus 10 gals. for each additional tub or shower, or dishwasher and/or clothes washer).

2. A water softener is fine if your water is too hard. You may wish to have your hot water alone (for washing) softened, or both your hot and some cold water, for washing and drinking but not for lawn watering and the like.

3. A boiler connection is handy if you have hot water or steam heat.

4. A furnace humidifier is good for health reasons, if you have warm-air heat.

5. A water conditioner is helpful if your water smells or tastes bad.

6. A sprinkling system allows you to take advantage of nighttime freedom of water use which may be restricted or at low pressure during the day.

## PLANNING A BATHROOM

Necessity as well as space should be considered. A very small bath will accomodate only one person at a time. If there is sufficient room to properly space the fixtures, a bath can accommodate two. However, if personal privacy is a consideration, the conventional type bathroom—no matter how large—will accommodate only one. But there are several modern methods of arranging the fixtures to accommodate two or more and still provide a measure of personal privacy.

The most economical arrangement is to have the three basic fixtures (lavatory, toilet and tub or shower), if all are to be used, side-by-side on one wall, or two on one wall and the third around the corner on an adjoining wall (see Figure 7–6). Placing one of them across the room or all three against separate walls may require reventing, and additional pipe installation costs.

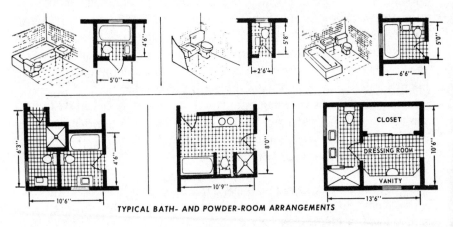

TYPICAL BATH- AND POWDER-ROOM ARRANGEMENTS

**Figure 7-6**

As far as possible, any bathroom should be planned for maximum convenience of the users. consider all of the following important details.

- A vanity or two to serve all your preparation needs.
- A shower enclosure for tub showering.

- A grab bar on the wall by the tub to help prevent a fall.
- A soap dish convenient to the tub.
- A towel rack convenient to the tub.
- A glass holder, toothbrush holder and soap dish convenient to the lavatory.
- A towel rack on or near the lavatory.
- Lighted mirror behind the lavatory.
- Electrical outlets convenient to the lavatory.
- A lighted medicine cabinet installed out of reach of small children.
- A paper holder by the toilet.
- Storage space for towels, paper, soap, etc.
- A built-in ceiling or wall heater, independent of furnace operation.
- A shelf convenient to the lavatory for sundries.
- Ample overhead light.
- Tiled walls and floor for easy cleaning.
- A pleasing, but not too distinctive color scheme. If you want to sell later, some prospects may not agree with your taste if it is too far out.

When ordering fixtures keep these thoughts in mind:

If the tub will be installed with both ends flush against walls, use the recessed type; otherwise use the corner type having one finished end. Be sure to order either a left or right hand tub, depending upon which end, when facing the tub, will have the connections.

In a toilet (water closet), low silhouette styling and exceptionally quiet flushing action are the most desirable features. Types include the traditional two-piece construction (water chest bolted to the back of the bowl), the newer lower, one-piece construction, and the wall-hung construction.

Lavatories may be wall-hung, with or without supporting legs, or may be cabinet (vanity) type having storage space below the basin. There also are drop-in lavatory basins for use with custom built cabinets. Two lavatories or vanities greatly increase the capacity of a bathroom to serve a family.

A separate shower cabinet takes little space and adds to bathing convenience, even when there is a tub.

All fixtures in a bathroom should be the same color, and should match or harmonize with walls and floor. White fixtures harmonize with any color.

For durability and easy cleaning every item that is part of a bathroom should be water and rust resistant. Ceramic tile, stainless steel or heavily chrome-plated fixtures, fittings and accessories are recommended. Walls may be enamel painted or waterproof papered, but plastic tile is better, and ceramic tile is best. The ceiling usually is enamel painted. The floor may be covered with asphalt, rubber, vinyl, linoleum or ceramic tile.

### PLANNING A KITCHEN

Some women prefer a small, compact kitchen in which few steps are required to reach any part; others like a big, roomy kitchen in which the family can gather. Most anyone likes a well-lighted kitchen that incorporates all possible conveniences. Your preference probably will depend upon whether you think of your kitchen simply as your personal workshop, or as a chief center of family activity. (See Figure 7–7.) Its size should coincide with your preference, and it should contain all you want from among the following conveniences:

- A single or dual basin, or two separate sinks. Dual basins provide dual facilities for food preparation and clean-up; two separate sinks accomplish the same, even better.
- A garbage disposer makes both preparation and clean-up much easier.
- A dishwasher, whether built-in or portable, is a great help.
- A refrigerator of ample capacity for family needs.
- A freezer of ample capacity.
- A large enough range with a range hood.
- A built-in oven and/or broiler.
- A ventilating fan.
- An appliance center with counter space and ample outlets.
- Ample countertop and or workable area.
- Ample storage space for foods, utensils, pans, dishes, etc.
- A planning desk with a phone extension.
- Efficient lighting.
- A handy trash disposal arrangement for paper, cans and garbage.
- A laundry corner if this is to be in the kitchen area.

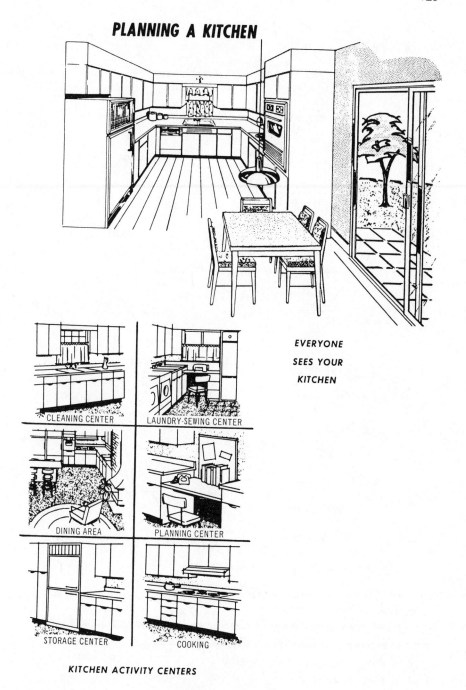

## PLANNING A KITCHEN

EVERYONE
SEES YOUR
KITCHEN

CLEANING CENTER

LAUNDRY-SEWING CENTER

DINING AREA

PLANNING CENTER

STORAGE CENTER

COOKING

KITCHEN ACTIVITY CENTERS

**Figure 7-7**

- A breakfast corner if this is to be in the kitchen.
- Tiled walls and floor.

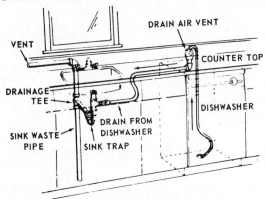

**TYPICAL SINK WITH DISHWASHER**

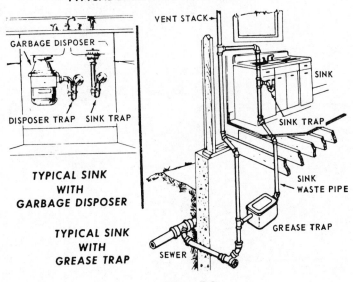

**Figure 7-8**

An attractive and efficient layout (see Figure 7–8) revolves around three basic work centers:

*First*, there is a storage center that includes the refrigerator, base and wall cabinets for packaged foods, and a countertop next to the opening side of the refrigerator door on which to set bundles and handle food. A freezer may also be included here, and a planning desk.

*Second*, there is the cleaning center, which must be handy to both other centers, and usually is the largest of the three. It centers around the sink, with ample counter space at each side. If used, the garbage disposer, dishwasher and trash disposal facilities should also be included here. And there should be generous storage in base and wall cabinet for utensils, cutters, glassware, dishes, soaps, towels, etc.

*Third*, there is the cooking center. This includes the range and oven and, again, ample counter space and storage space for pots, pans, seasonings and cooking utensils. If large enough, the counter space between this center and the cleaning center is an excellent location for appliances such as a toaster, mixing bowl, blender, food chopper, etc. Should there be insufficient counter space, an island, peninsula or even a table arranged to abut upon this center can be used instead.

See Figure 7–8 for sink connection diagrams to appliances and the sewer.

The layout of a kitchen depends less upon its size than upon the placements of windows and doors. However many doors there are, the wall space they occupy must be counted out. Window wall space cannot be shared with an upright refrigerator or freezer, a built-in oven or wall cabinet, but can be shared by a sink, counter, breakfast nook, etc., that stands lower than the windowsill. There are four widely accepted general arrangements of the three work centers in a kitchen: U-shape, L-shape, Corridor and One-wall (See Figure 7–9). One of these can be used for any kitchen area that is not extremely unusual.

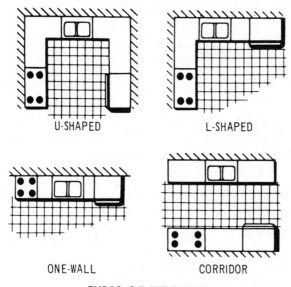

U-SHAPED    L-SHAPED

ONE-WALL    CORRIDOR

*TYPES OF KITCHENS*

**Figure 7-9**

**LOOKING AHEAD**

A growing family may be faced with the need for more space at a time when moving to a larger house is less desirable (or less economically feasible) than adding on to the existing house would be. If practical, the addition of just one bedroom and bath, or even a half bath, might solve the problem to best advantage all around. Or, possibly, the addition of some conveniences not originally included—a dishwasher, laundry room, basement shower or half bath, sprinkler system, etc., may prove a happier solution than seeking or building a new house in another neighborhood to obtain them. Then, again, should you sell your house, quite possibly the prospective purchasers will be influenced by these same thoughts.

When building a home it is wise to consider such future needs. A hurried decision to conserve on plumbing at the expense of sacrificing

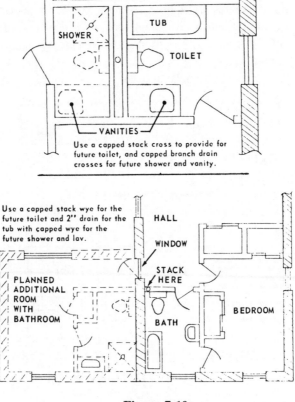

**Figure 7-10**

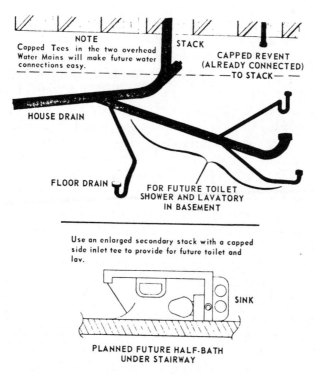

NOTE
Capped Tees in the two overhead Water Mains will make future water connections easy.

STACK

CAPPED REVENT
(ALREADY CONNECTED)
—TO STACK— —

HOUSE DRAIN

FLOOR DRAIN

FOR FUTURE TOILET
SHOWER AND LAVATORY
IN BASEMENT

Use an enlarged secondary stack with a capped side inlet tee to provide for future toilet and lav.

SINK

PLANNED FUTURE HALF-BATH
UNDER STAIRWAY

**Figure 7-10 (cont.)**

tomorrow's requirements can prove to be very costly later on. The cost of planning now for later possible additions is negligible compared to the cost of opening walls and floors to later install additional piping and connections.

The first step in future planning is to install piping of adequate size for future needs, instead of the minimum size immediately needed. For instance, with respect to drainage this might mean using a main stack, instead of a secondary one, to serve a sink to allow for later addition of a bath or powder room that will use this stack. Or, it might mean providing a 2″ branch drain from the sink instead of 1½″ size to allow for addition of a dishwasher. With respect to the water supply it could mean installing the largest diameter building main allowed instead of the smallest, and/or installing a ¾″ cold-water branch where a sprinkler system can later be connected to it, instead of the usual ½″ size.

See Figure 7–10 for diagrams which show practical planning methods.

The next step is to stub-in necessary connections to the drainage and water-supply systems where these will later be needed. Breaking into a stack to connect a branch drain is greatly simplified if there is a plugged tee already there; and, to a lesser extent, this applies to a water line. For instance, a large closet back-to-back with the toilet in a bathroom can be stubbed-in for conversion to a powder room. To do this, you would install a sanitary cross instead of a tee to serve the bathroom toilet, leaving the opposite opening for the future powder-room toilet. This opening could be plugged or, better still, the toilet closet bend could be installed in it and be capped where it emerges through the closet floor. In similar manner, crosses or extra tees can be installed to the stack to provide for future lavatory waste and, if required, revent connections, and can be plugged at the stack or can be extended to where they will be needed through the closet wall, and be capped at this point. And the same applies to the necessary water-supply connections. If plugging is done inside the walls, future conversion to a powder room will necessitate breaking into the walls, but the pipes themselves will not have to be disturbed. If capped branches are installed to project through the closet floor and walls, no wall breaking will be required.

Similar stub-ins which also would save future costs are: for a dishwasher, for an attic bathroom, for a powder room on the first floor below a second floor bathroom, for a basement bathroom, half-bath or shower, for a shower at one corner of a large bathroom, for an extra kitchen sink, for laundry equipment, for a water softener, for a running-water cooler, etc. If you look at your house plan with an eye to the future you will probably note many improvements of this nature which may later be desirable. You can even stub-in the connections needed for a whole additional stack, to serve a planned addition onto the side of the house.

*Note*: Whenever stub-in provisions are made in new house construction, these should be indicated on the house plans and filed away for future reference.

## PLANNING THE DRAINAGE SYSTEM WITHIN THE HOME

Please refer to Chapter 6 for full information on working with pipe and tubing; your first task will be to select the type of pipe and stack size. There are four basic types of house systems in general use—your local code mey specify one or more of these as permitted for use in your area.

1. The oldest and most universally specified is the hub-and-spigot cast-iron and steel pipe system. This uses cast-iron pipes and fittings for each main stack, the branches to toilets, the house drain, and any other branch drains, such as the basement floor drain, which must be underground. The cast-iron is available in

2, 3 and 4″ diameter sizes. Steel pipe and fittings in 1½″ or 2″ diameter size are used for all other branch drains and for all vent lines connected to a stack. Either cast-iron or steel may be used for secondary stacks. This is a durable system, but is much the heaviest and most laborious to install.

2. Newly developed for eventual replacement of the above is the hubless cast-iron and steel pipe system. This is similar to the above in all respects except that the cast-iron parts are more easily assembled and, due to the lack of hubs, occupy less space within a partition or wall.
3. Also long preferred in some localities is the copper pipe system. All above-ground parts are of copper fittings and pipes (in 3″ diameter size for stacks, toilet branches and house drain, and in 1½″ diameter size for all else). A 3″ copper stack has the same waste flow capacity as one of 4″ cast-iron and requires less partition space, is somewhat lighter and there are fewer joints to make.

Figure 7–11 shows drainage-pipe sizes you should use, and the space required for a vertical run of pipe inside of a wall or a partition.

**DRAINAGE-PIPE SIZES**

| | |
|---|---|
| •MAIN STACK | 3″ or 4″ |
| •SECONDARY STACK | 1½″ or 2″ |
| •HOUSE DRAIN | Like Stack |
| •HOUSE SEWER | 4″ |
| •TOILET BR. DRAIN | Like Stack |
| SEPARATE TOILET VENT | 1½″ |
| •LAVATORY BR. DRAIN | 1½″ |
| LAVATORY VENT | 1½″ |
| •TUB OR SHOWER BR. DRAIN | 1½″ |
| TUB OR SHOWER VENT | 1½″ |
| •SINK BR. DRAIN | 1½″ |
| •GARBAGE-DISPOSER BR. DRAIN | 1½″ |
| •DISHWASHER BR. DRAIN | 1½″ |
| SINK VENT | 1½″ |
| •LAUNDRY-TUB or AUTO. WASHER BR. DRAIN | 1½″ |
| TUB or WASHER VENT | 1½″ |
| •BASEMENT DRAIN | 2″ |
| •ANY OTHER DRAIN (If Code Permits Connecting to House Sewer) | 2″ or 4″ |

•These Lines Must Be Assembled With Sanitary Drainage Fittings – Not Standard-Pipe Fittings. Others Use Standard Fittings.

**SPACE REQUIRED FOR A VERTICAL RUN OF PIPE**

(Inside a Wall or a Partition)

| PIPE SIZE | CAST-IRON PIPE AND FITTINGS | STEEL PIPE | | COPPER PIPE | | PLASTIC PIPE | | NO-HUB C-I PIPE | |
|---|---|---|---|---|---|---|---|---|---|
| | | PIPE ONLY | W FITTINGS | PIPE ONLY | W FITTINGS | PIPE ONLY | W FITTINGS | PIPE ONLY | W FITTINGS |
| 1½″ | – | 2″ | 3″ | 1¾″ | 2″ | 2″ | 3″ | 2″ | 3″ |
| 2″ | 4″ | 2½″ | 3½″ | – | – | 2½″ | 3½″ | 2½″ | 3½″ |
| 3″ | 5¼″ | – | – | 3¼″ | 3½″ | 3½″ | 4½″ | 3½″ | 4½″ |
| 4″ | 6¼″ | – | – | – | – | – | – | – | – |

**Figure 7-11**

Pipes, cut to exact lengths on the spot, and fittings are soldered together. Lighter work, and no more skill, is required than to assemble cast-iron and steel. Parts that are underground or buried in concrete are, however, of cast-iron (as above). See Figure 7–12 for a plan for a house drain and a house sewer that should suit your needs.

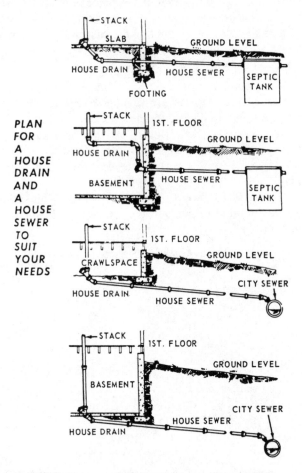

Figure 7-12

4. Newest and best from every standpoint (though not yet universally recognized by plumbing codes) is the plastic pipe system. Parts are all much the same as in a copper-pipe system (1½″ and 3″ diameter sizes), but are much lighter and easier to install. In fact, watertight joints are made by using a solvent that is simply applied by brush—anyone can make them. Tough, heat-resistant, and less corrosive than either metal, this system is exceptionally

durable and trouble free. Plastic may be used for all parts of the system—even, in some cases, underground or in concrete.

## HOW TO EXTIMATE YOUR NEEDS

Among the installations shown here you probably will find one or more drainage systems for house plans like—or nearly like—yours. If one of the systems exactly fits your plans you can use the same materials listed for this system, in the type of pipe you have chosen. Should one of them almost, but not quite fit your plans, you can easily estimate the additional parts (usually, just pipe lengths) needed to adopt it. If your house plans call for two or more stacks, you will need two or more of the typical systems.

## PLANNING THE WATER SYSTEM—
## FIRST SELECT THE TYPE OF PIPE

See Chapter 6 for details on working with the pipe of your choice. There are two types of water-line parts:

1. Galvanized steel pipe and fittings, used in a great many homes, especially where economy is important. All pipe lengths are outside threaded and all fittings are inside threaded, and are assembled by being screwed together. This pipe should not be buried underground, nor in concrete.
2. Copper tubing and fittings, considered the finest and longest lasting for a water-supply piping installation, is available in both rigid and flexible types, but the rigid is recommended for new home installations. Assembly is accomplished by solder joining pipes into fittings. This pipe may be used anywhere. Sizes commonly used are for mains (¾"), for branch lines (½").

In addition to pipe and fittings every domestic installation requires some valves. A first-class system contains a number of valves to shut off water flow to various parts of the system so that one part can be shut down for repairs or alterations without affecting other parts. The type generally used is called a Stop Valve, though other types for special purposes are available. (See Chapter 6.) Threaded brass valves are used with steel pipe, solder-type copper valves with copper pipe.

## HOW TO ESTIMATE YOUR NEEDS

A typical water-supply system for one bath and one kitchen is shown in Figure 7–13. From this you can determine the kinds of fittings re-

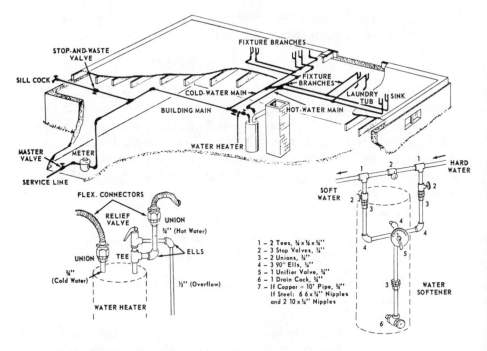

**A TYPICAL WATER-SUPPLY SYSTEM FOR ONE BATHROOM AND A KITCHEN**

### Figure 7-13

quired, and can approximately estimate the number of each you will need. To determine the approximate footage of pipe required use an elevation plan of your house from the front and another from the side. Draw lines on these to indicate where the horizontal and vertical pipe runs will be, then measure in scale the lengths of these lines to arrive at a total for each pipe size.

### ORDERING REQUIREMENTS—PIPES AND FITTINGS

On the basis of your estimates of drainage and water-supply pipe and fittings make a list of parts. In the case of drainage, simply list one or more basic packages. Arrange the parts in your list according to kinds of pipe, and specify each kind (cast iron, copper, etc.). Be sure to specify sizes (diameters) for pipe and, in the case of fittings, list these by their proper designations. See Chapter 6 for help in identifying fittings.

If in doubt as to your actual requirements be generous in preparing your list; check with your supplier to be sure he will accept for return and full credit all unused fittings and uncut pipe lengths—Wards, for example, has this policy.

A hot water heater and water softener or conditioner should be selected, if needed, and the proper fittings chosen at the same time. Water that is too hard makes washing difficult, and tends to clog the water system with mineral deposits which accumulate in the lines. Water tainted with iron, sulphur, or other minerals may have an offensive odor and taste. If your water supply is affected by either of these conditions, you will need a softener or conditioner. Figure 7–13 illustrates the fittings normally needed for these fixtures.

## SELECT PLUMBING FIXTURES, FITTINGS AND TRIM

The kind, styling, finish and coloring of fixtures is a matter of personal preference. Choose a fixture in keeping with your requirements, and remember that when ordering them the fittings and trim will also be needed.

Fittings are the faucets, filler spouts, strainers and/or drains that are mounted in or on such fixtures as tubs, showers and sinks. Some fixtures are offered with fittings, but most are not since there are a variety of fitting styles, finishes, etc. *Trim* refers to the external (usually, in sight) plumbing used to connect each fixture with the building drainage and/or water-supply systems. The trim (pipe and fittings) usually is either brass or chrome plated. A toilet requires a water-supply pipe; a lavatory or a sink requires two water-supply pipes (hot and cold) and drainage fittings usually comprised of a trap and tailpiece and/or a drain extension. (Connections to tubs and showers generally do not require trim since they are made behind walls, with ordinary pipe and fittings.)

Water-supply trim is available pre-assembled into one piece ready for connection to the faucet and to the in-wall (or floor) nipple of the building supply system. Both stiff and flexible types, with or without integral shut-off valves, and in various lengths are offered. Drainage trim may be offered in single pieces or in groups of pieces intended for use with a lavatory or a single or double basin sink. Make your selections and list these, also, when you order your fixtures.

## HOW TO PLAN A MODERNIZATION

The fixtures in an old house are quite often much less attractive and much less convenient than the new. Old fixtures, even though still usable, will date a house and make it less comfortable to live in, and less saleable.

If you have such fixtures you can add greatly to your own pleasure and the value of your house by replacing them. See Figure 7–14.

**THE OLD**

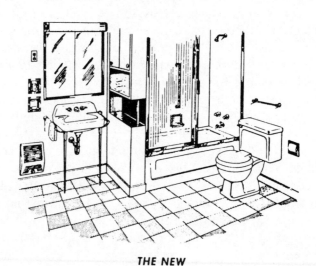

**THE NEW**

**Figure 7-14**

In many cases, too, new fixtures will also increase the capacity of your plumbing system to serve you. For instance, a new tub with built-on shower will make bathing quicker, serve more people during a rush period, than a showerless tub. A lavatory set in a modern vanity cabinet can double as a dressing table and save many steps. Or, a double sink with a garbage disposer, a dish spray and a convenient size drainboard will speed the after-dinner cleanup.

Generally, new fixtures can be connected to old house plumbing without need of opening walls to alter the in-wall or in-floor pipes. In such case all you need order in addition to each new fixture with fittings is the necessary trim. Even when, as may sometimes be the case, an old sink branch drain proves to be undersized for a dishwasher, it often can be replaced without opening the wall or floor.

## ADDING A POWDER ROOM

If you have a two-story home with an upstairs bath only, a first-floor powder room will be more than worth the relatively small cost of installing it. Or, if yours is a one-story home with living space in the basement, this also applies to a new basement powder room. Running water pipes to new fixtures are seldom a problem since these pass easily through practically any partition and can be run up, down or sideways as need be. Drainage, alone, presents problems.

Adding a basement powder room usually is quite simple, if you choose a location at the foot of the main stack. In such case, the exposed portions of stack in basement and attic can be broken into for drainage and for venting—there's no need to break through walls or dig up the floor. Adding one on the first floor also can be simple if you can choose a space (under a stairway, where a closet was, or the partitioned-off part of a large room) where the toilet can be backed up to the existing main stack. Even when the existing stack can't be used, there are methods of installing the required new stack which avoid opening of walls.

## ADDING THE SECOND BATH

If space can be found, adding a second bath can be accomplished in the same manner as adding a powder room. Quite often, however, a second bath—together with additional bedrooms—is added as an extension or wing onto the building. When the addition is so located that a new house drain can be installed to join the existing house sewer out in the yard (so that it is necessary only to trench the yard), this is the simplest method. However, if the location is such that a basement lies between the new bathroom and the house sewer, it will have to connect either with the existing house drain or with the main stack.

To connect it with the house drain means trenching through the basement floor and the foundation on the side where the bath is being added. A better method is to use a (new) suspended house drain which will reach from the foot of the new stack to the original main stack, and to join this with a wye or tee to the main stack. See Figure 7–15. The

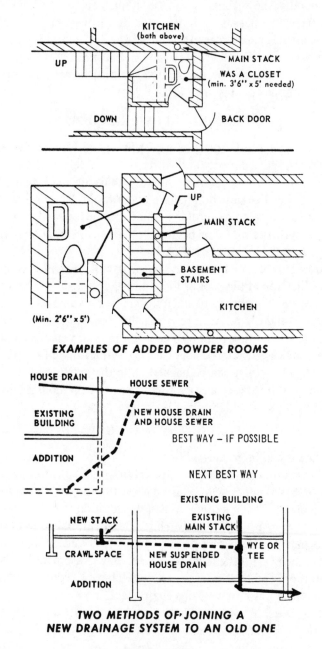

**EXAMPLES OF ADDED POWDER ROOMS**

**TWO METHODS OF JOINING A
NEW DRAINAGE SYSTEM TO AN OLD ONE**

**Figure 7-15**

suspended house drain will necessarily be exposed in the original basement, so prefereably run it as close to a basement wall as possible (where it can run below the joists without detracting from headroom).

PIPE CONCEALED IN A SOFFIT

**TYPICAL METHODS OF CONCEALING STACK AND HORIZONTAL PIPES**

**Figure 7-16**

If a new bath is to be on the second floor above, for instance, an enclosed back or side porch, it is best to run the stack down in or outside a wall of the porch—then use a suspended house drain. The boxing around the cleanout at the foot of the stack should contain a removable panel for access, when required. See Figure 7–16 for typical methods of concealing stack and horizontal pipes.

When adding fixtures—except for toilets—lavatories, tubs, showers and laundry facilities or dishwashers can be drained through 1½" or 2" and can be vented, if needed, through 1½" pipes. Pipes of these sizes in copper or plastic can nearly always be run through existing partitions. See Figure 7–17 for plans and minimum spacing requirements.

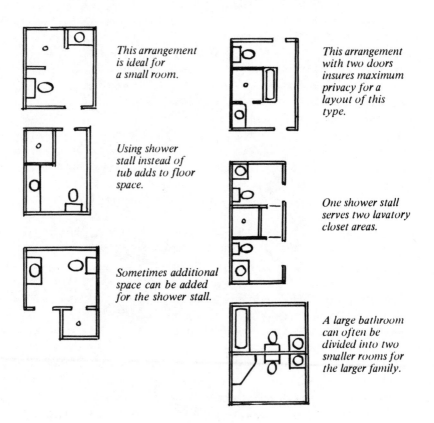

*This arrangement is ideal for a small room.*

*This arrangement with two doors insures maximum privacy for a layout of this type.*

*Using shower stall instead of tub adds to floor space.*

*One shower stall serves two lavatory closet areas.*

*Sometimes additional space can be added for the shower stall.*

*A large bathroom can often be divided into two smaller rooms for the larger family.*

**Figure 7-17**

# 8

# INSTALLATION OF PLUMBING

When making pipe runs through walls and floors (Figure 8–1), floor and stud spaces required for various pipes will have to be taken into account. See Figure 7–11 for measurement details. In new construction the only problem is to plan sufficient space for any stack or closet bends—other pipes will fit in normal-sized wall and floor spaces. For a remodeling job it may be necessary to furr out existing partition studs as shown in Figure 8–1, or to box in a new stack as also shown. The joint space between a floor and ceiling is practically always sufficient for underfloor piping.

## HORIZONTAL RUNS

Pipes that must run across a partition somewhere between floor and ceiling will have to cut through the partition studs. Since the largest of such pipes usually is 1½" size (not over 2"), notching the studs does not present a serious problem. To avoid weakening them unduly, do not notch deeper or taller than necessary. Then, after the pipe is in place, face each notch with a reinforcing strip of steel, as shown.

Pipes that must cross through floor joists present more of a problem. No joist should be notched more than one-fourth of its depth, and this only near its ends (not in the center half). Deep notching can be avoided to some extent by taking advantage of the flooring by letting the top of the pipe extend up through the sub-flooring and even, if necessary, by adding furring strips on top of the joints or sub-flooring as in Figure 8–1. Should it be absolutely necessary to notch a joist improperly, it should be reinforced with at least a 2 × 4 on one side, as shown—preferably, on both sides. If a closet bend must cross a joist (and never more than one),

# Pipe Runs Through Walls and Floors

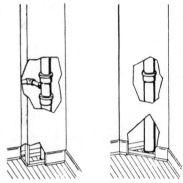

**BOXED-IN AND IN-CORNER STACKS**

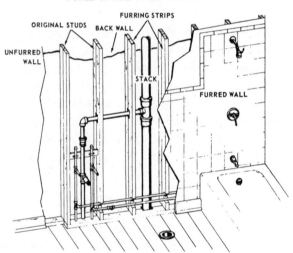

**INCREASING WALL DEPTH TO COVER PIPES**

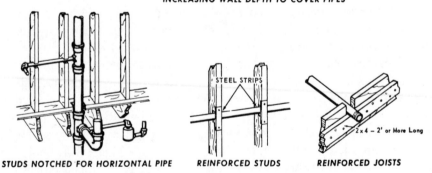

STUDS NOTCHED FOR HORIZONTAL PIPE      REINFORCED STUDS      REINFORCED JOISTS

Figure 8-1

notching and reinforcing can't be done. The joist must be cut, then its ends supported with double headers as in Figure 8–2.

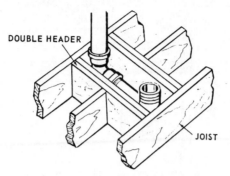

**HEADERS REINFORCE A CUT JOIST**

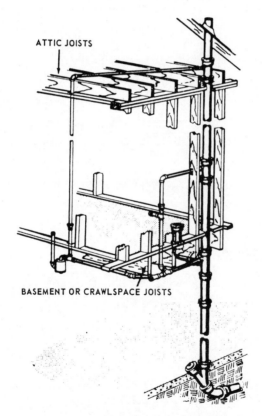

**HORIZONTAL PIPES ABOVE AND BELOW JOISTS**

**Figure 8-2**

Whenever possible locate horizontal runs in floors or ceilings between the joists (see Figure 8–3). Or, if practical, as in an attic or
crawlspace, locate them above or below the joists.

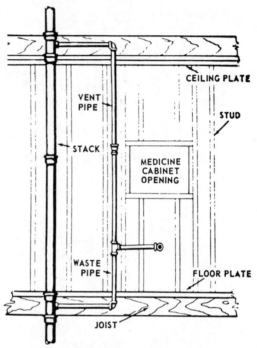

**HORIZONTAL PIPES IN JOIST SPACES**

**Figure 8-3**

## STAGGERING A STACK

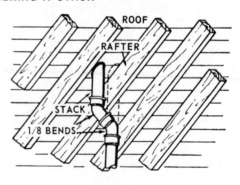

**Figure 8-4**

To stagger a stack, see Figure 8–4. Sometimes a joist or an offset partition on the floor above will prevent running a stack straight up. If so, it can be staggered by using two ⅛ bends, or two ¼ bends as shown in Figure 8–4. Wherever a stack is located in an outer wall it is customary to so stagger it at the top so that it will pass through the roof in from the wall rather than right at the roof edge.

### BEFORE BEGINNING PIPE ASSEMBLAGE

First check your shipment against the shipping list to be sure you have all parts. A help in checking the parts will be the illustrations in Chapter 6. Unpack the cartons carefully and examine all the packing for small parts before disposing of the cartons and packing. Sort the parts in matching, separate groups by pipe sizes, to save time later, and store them out of the way of other construction work. For protection, store the fixtures in their own cartons until actually needed for installation.

Before beginning an installation, plan ahead to its completion. If you have a set of architect's plans or your own sketches, study these to acquaint yourself with the necessary details. Obtain an overall understanding of the installation you will make.

*Note*: Detailed plans involving every pipe and fitting are not actually needed. Piping runs can be designed on the job if you know where the fixtures are to be, and that they will fit in, and that partition and floor spaces are (or can be) made available for the stacks and other piping required to reach them. Also, the house drains can be located as required by the stacks, and be run out to join the house sewer, and thence to line disposal. The hot-water heater can be located as required to properly vent it, and it and the two water mains can be conveniently connected to the building main, run in from your source of water supply.

### PROCEED IN THIS SEQUENCE

If you are remodeling, the following sequence of steps is recommended. It also is the most practical for a new house, unless the number of men on the job or other construction details make it advisable to proceed differently.

1. Install underground house and other drains (Figure 8–5) or a suspended house drain (Figure 8–8).
2. Do fixture roughing-in (Figure 8–10).
3. Install stack (Figure 8–13), also second stack (Figure 8–17).
4. Install the branch drains and the vent lines (Figure 8–20).

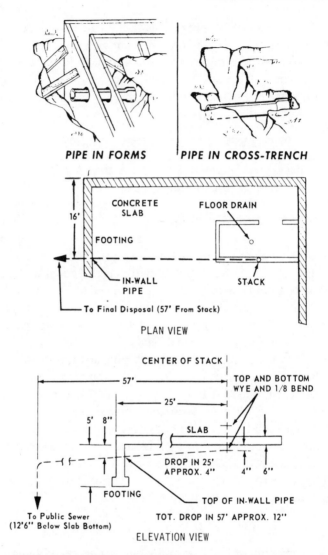

**PIPE IN FORMS** | **PIPE IN CROSS-TRENCH**

PLAN VIEW

ELEVATION VIEW

**TYPICAL PLANNING FOR LOCATION OF AN IN-WALL PIPE**

**Figure 8-5**

5. Install the water-supply piping (Figure 8–28).

6. Install fixtures (Figures 8–36 and 37).

The above completes the in-house installation work. For exterior work involving the house sewer and a private final disposal (if used), refer to Chapter 9.

### INSTALLING UNDERGROUND HOUSE AND OTHER DRAINS

With new construction much work is saved by locating and positioning any pipes which must pass through the foundation, prior to pouring it. (One that will pass under a foundation can be trenched in after foundation is poured, at the time the house drain is installed.) And there would be no point to pouring a basement floor until after all the underfloor piping is located and installed. Cast-iron generally is used for piping of this type, though plastic may, in some cases, be acceptable. Copper is not.

To install an in-wall pipe, and to locate a house-drain pipe that will pass through a foundation wall or footing it is first necessary to know the depth of the public sewer (or elevation of private septic tank location), and the distance from sewer (or tank) to the wall. In the case of a public sewer the exact location of the sewer is at the point where your house sewer will empty into it and this information should be available from the local facility. A survey may be required to spot a septic tank location. In either case, a plan will be required to locate the in-wall pipe, as in Figure 8–5.

When the foundation forms are erected, spot the location on an inner form. If practical, cut through forms (or, if a trench is being used, cut a cross-trench) and install a length of pipe, hub to the inside, as shown in Figure 8–5. If this isn't practical, bolt a sleeve or a dowel inside the forms at the location. A sleeve will remain in the finished wall and must be large enough inside to pass the 4″ pipe through, and strong enough to hold up when the concrete is poured. A dowel will be chiseled out to make room for the pipe so it should be about the pipe diameter, and can be made of wood or a short piece of 4″ tile pipe. In either case, after the foundation is poured, any opening around the pipe in the wall should be filled from inside and outside with hydraulic cement, if it is desired to have it water-tight.

*Note*: Although a water-pipe hole can be driven through a poured foundation without too much trouble, such practice often leads to a ground-water leak into basement areas unless the space around the pipe is packed inside and out with a hydraulic cement. Therefore, if practical, it also is advisable to install a length of Type K copper (¾″) building supply pipe through the forms prior to pouring.

### INSTALLING THE HOUSE DRAIN

After foundations are in, and before the basement floor (or a basementless house slab) is poured, trench the ground and install your house drain or drains joining it to the in-wall pipe and extending it to the wye which will project up through the basement floor or slab at the foot of the stack. Also do install any basement and/or other drains to be buried under the floor or slab, connecting these to the house drain. Even if your house will have a crawlspace instead, it will be easier to install the house drain and wye before the first-floor construction begins and the space is reduced to one that is too cramped for easy accessibility.

A basement floor should not be poured until after the roof is on the building. It is therefore a simple matter to determine a stack location in the building to prepare the first-floor opening for the stack, then to use a plumb bob as seen in Figure 8–6 to spot the foot of the stack at the basement level. This also applies if you are remodeling an existing building. To spot the stack foot before a slab is poured or before the building is

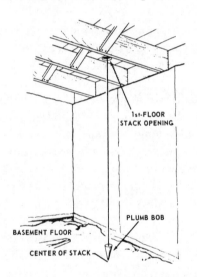

LOCATING STACK BASE

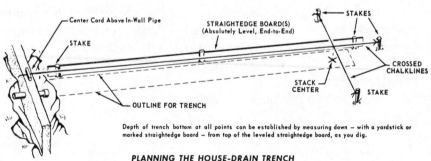

Depth of trench bottom at all points can be established by measuring down — with a yardstick or marked straightedge board — from top of the leveled straightedge board, as you dig.

PLANNING THE HOUSE-DRAIN TRENCH

**Figure 8-6**

erected over a crawlspace does, on the other hand, require measurements plotted from a building plan. Lay out enough of your first-floor plan on the ground to properly locate the stack foot. In any case, when the stack foot is located, drive stakes into the ground as in Figure 8–6 (planning the house-drain trench)—each two to four feet from the stack foot location—and stretch two chalklines between them to cross at the exact center of the stack foot. One of these lines should extend from the in-wall pipe, and will mark the centerline of the house-drain trench. This drain should run as straight as possible to the foundation, and thence (as the house sewer) on out to the final disposal. If not straight, ⅛ or ¼ bends must be used. Curving it, even slightly, without using bends is not good.

*Note*: A house drain or sewer can turn sideways using ⅛ bends but, preferably not ¼ bends, or can turn down to a lower level using ⅛ or ¼ bends. If the pitch of any section must exceed 1″ per 4′, then this section should be sloped at least 45°, preferably, up to 90°.

Dig the trench to proper depth and pitch, keeping in mind that pitch should be 1″ per 5′ (maximum, 1″ per 4′) from the foot of the stack all the way to the final disposal. Any lesser pitch will cause water to flow fast enough to leave some greases behind. An easy way to determine the correct pitch as you dig is illustrated in Figure 8–6 and 7. Should you dig too deep at any point, refill to the proper level and tamp the dirt in very firmly. It is important that the trench bottom be hard packed, not loose fill, since any later settling may cause the drain run to sag. Figure 8–7 shows the methods of aligning the cleanout-wye, various cleanout arrangements and an assembled house drain.

Begin the house drain by pre-assembling the wye, ⅛ bend and a length of pipe, either standing them vertically, or laying them out horizontally squared in similar fashion against the base of the foundation (squaring is essential to obtain correct angle between wye, bend and pipe). Before making this assembly, however, make certain that the two top openings of the wye will be above the basement floor level, when the floor is poured. If the trench, at the point under the stack, is so deep that they will not be, it will be necessary for you to install a sufficient length of cut pipe between the wye and the ⅛ bend (or ¼ bend) to assure having the wye reach to the above floor level.

If the drain will pass under the foundation and there is no in-wall pipe, you can lay the above assembly in the trench and join additional pipes to it as needed to complete the house drain. If there is an in-wall pipe, lay the assembly in the trench and loosely lay in enough pipe to reach the in-wall pipe—one of the pipe lengths most likely will have to be

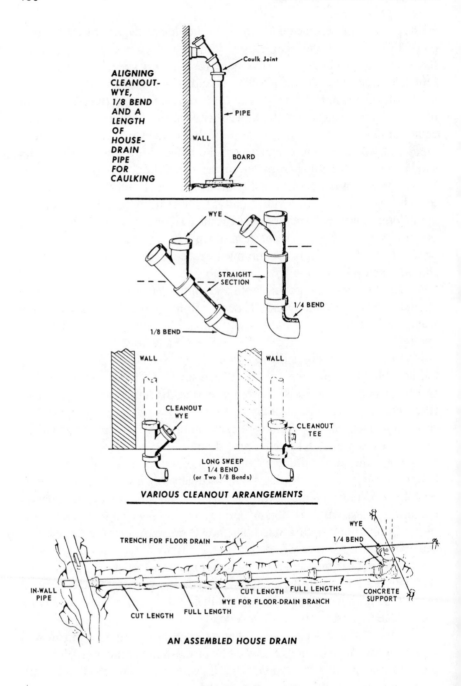

Figure 8-7

cut to fit. After thus assuring a proper length of run, start at the in-wall pipe and join all pieces ending with the vertical portion of wye exactly vertical and centered at the center of the stack foot. Support it accurately in this position. Bricks, rocks or 2 × 4s can be used, but the best support will be some concrete wedged (before it is set) around the bend and the base of the wye.

## INSTALLING HOUSE-DRAIN BRANCHES

If a second stack, a basement drain, or any other branch drain (as from a basement laundry tub) is to join the house drain, the wye or tee for such a drain must be incorporated into the house drain as it is assembled. In the case of the stack, a wye can be positioned in the house drain at the foot of the stack—like the first stack wye—or, if the stack is not directly above the house drain, it can be connected to it by a horizontal branch. Any horizontal branch can be joined to the house drain by a tee or a wye—which must be positioned (rotated) so that the branch will join the house drain at the same level, or from any higher level at which it can be joined to the wye or tee by a ½ bend or a ⅛ bend. To trench-in and assemble any second-stack wye or any branch drain, use the same method as for a house drain that must join an in-wall pipe. That is, don't join all pieces until the run has been loosely assembled in entirety—then do the final joining from the junction back to where the drain originates.

Don't fill in the trenches until after the drainage installation has been completed and tested, unless you feel that testing will be unnecessary. When you do fill in, tamp the dirt firmly all around and above the pipes—before pouring the basement floor (which should slope about 1" per 5 to 10′ from all sides into the basement drain. Note that the completion of the run from beyond the foundation to the final disposal is discussed in Chapter 9.

## INSTALLING A SUSPENDED HOUSE DRAIN

Note a suspended drain, out to the first length of pipe in the house sewer, generally is constructed of the same type of pipe (cast-iron, copper or plastic) used for the stacks. Any in-wall pipe required can be installed with less labor at the time the foundation or the basement wall is poured; but the remainder of the installation is best done after the roof is on the house and the roughing-in has been accomplished. Figure 8–8 shows the types of suspended house drains and methods of installing the in-wall pipe.

# Installing a Suspended House Drain

### NOTE

A suspended drain, out to the first length of pipe in the house sewer, generally is constructed of the same type of pipe (cast-iron, copper or plastic) used for the stack(s). Any in-wall pipe required can be installed with less labor at time foundation (or basement wall) is poured; but the remainder of the installation is best done after the roof is on the house and roughing-in (refer to next section) has been accomplished.

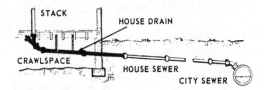

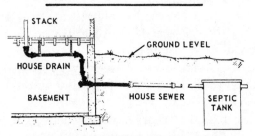

**TYPES OF SUSPENDED HOUSE DRAINS**

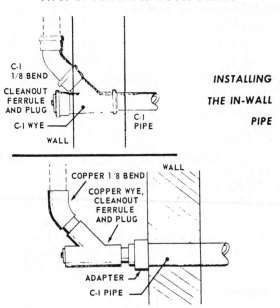

**Figure 8-8**

When a suspended drain is used, if the house drain will be above ground underneath a house built over a crawlspace, it is a suspended drain. A suspended drain also may be installed in a basement, if burying the drain beneath the basement floor would place it too low for proper connection with a public sewer—or, the case of a private final disposal, would require burying the septic tank excessively deep in the ground.

When installing an in-wall pipe, when the drain is in a crawlspace it usually will pass through a foundation wall or footing on its way to the final disposal. In such a case, the in-wall pipe length can be planned and installed exactly as explained preceding for an underground drain.

Note that whatever the type of system pipe you have, the cast-iron should be used for in-wall and/or underground portions. When the drain is in a basement it will pass out through a basement wall—somewhere between the basement floor and ceiling. (See Figure 8–8). The in-wall pipe will be a wye joined to a length of pipe, usually installed as seen in Figure 8–8. These two pieces must be properly joined, sealed and caulked before embedding them in the wall. As with an underground drain, they are positioned in the wall at a point on a line between the stack foot and the final disposal, and at a height which will allow the house sewer (the underground portion of the run) to drain correctly into the final disposal. Whatever this height (in the wall) may be, the basement portion of the run can be arranged accordingly.

Note that the wye generally is installed in the wall so that the basement portion of the drain can rise vertically as close to the wall as possible, then run horizontally to the stack at an elevation just beneath the first-floor joists (or between two of them, if practicable). This way the drain does not occupy basement space unnecessarily.

When any pipe passes through a basement wall above the floor level the opening around the pipe must be tightly sealed, or ground water will seep into the basement. If necessary, fill around the pipe from inside and outside with hydraulic cement.

### INSTALLING THE SUSPENDED DRAIN

Figure 8–9 shows typical ways of making bathroom connections to a suspended house drain, sink only connections to a suspended drain, and the general installation of a suspended drain. If the stack serves a first-floor bathroom, provision for the toilet, tub and lavatory branches must be made when assembling the suspended house drain. Several of these are seen in Figure 8–9. Note that when a suspended drain is used a first-floor toilet need not be close to the stack—it can empty directly into the

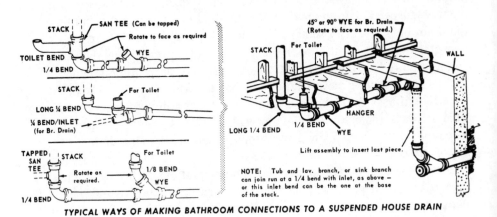

**TYPICAL WAYS OF MAKING BATHROOM CONNECTIONS TO A SUSPENDED HOUSE DRAIN**

**TYPICAL SINK (ONLY) CONNECTIONS TO
A SUSPENDED DRAIN**

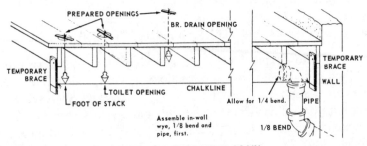

**INSTALLATION OF A SUSPENDED DRAIN**

**Figure 8-9**

suspended drain at any convenient point. So can other branch drains if proper reventing is provided.

If the stack serves a second-floor bathroom with a kitchen on the first floor, the only provision which may be necessary in the house drain will be for the sink. Again, see Figure 8–9.

In any event, begin by assembling the vertical portion of the suspended drain, from the in-wall wye or pipe up to but not including the ¼ bend at the top of this portion. Complete each joint in turn. When cutting the vertical pipe length be sure to allow room for the ¼ bend that will be on top of it.

Next, locate the foot of the stack and make the opening for it through the floor. Also, locate and make openings for the closet bend (or the short pipe to serve the toilet), the tub and/or lavatory branch drains, and the sink branch drain, as required. Refer to the section on roughing-in.

With all openings made, hang plumb bobs (or equivalent weights) through them, then stretch a chalkline (under the joists) from the center of the stack foot to the top center of the assembled vertical section. The chalkline will mark the path of the horizontal portion of the suspended drain. Together with the hanging plumb bobs it will provide a means of locating the various fittings required in the portion of the run.

Make the measurements carefully with allowances for the necessary fittings, then cut the pipe lengths required. Pre-assemble (on the floor) as many of the pieces as you can handle and lift into place. If the entire horizontal portion is thus pre-assembled (which it probably can be if using copper or plastic), be sure that the ¼ bend at the stack foot and the ¼ bend at the top of the vertical portion (and, also, any other fittings) are pointed correctly to later join the pipes they must join. Use the methods shown in Figure 8–7 for squaring these fittings. Starting at the assembled vertical portion, if you have two or more pre-assembled horizontal portions, lift the pre-assembled portions into place and temporarily secure with stout wires to joists above. Check to see that a ¼ bend at the foot of the stack is properly located. Also, check the pitch of the entire run. When satisfied, use permanent steel pipe hangers to securely hold the entire run—to the joists above—then complete the remaining joints.

## ROUGHING-IN FIXTURES

The exact location of all fixtures must be established before the stack that serves them can be assembled with fittings or the fixture branch drains are properly positioned. This is called roughing-in. In a new house it is accomplished after the roof is on, but before the floor and partitions are finished. When remodeling it is good to start the installation with this step, and to accomplish it, preferably, after the walls and floor have been stripped, if new flooring and/or wall will be installed.

Since fixtures are not made to uniform sizes or specifications it is necessary to use the exact measurements of each individual fixture. Check the dimensions and roughing-in specifications (size and placement of each inlet and outlet) of each of your fixtures, either by making actual measurements or by referring to the fixture owner's guide. Using these dimensions and specifications, plot the position of each fixture and mark the locations of supply and drain pipes to be installed in walls and/or floor. Prepare any openings, notches, etc., required for the pipes.

**For a lavatory**

A wall-hung lavatory (see Figure 8–10) is suspended by hangers which, in turn, must be securely fastened to the wall. A 1 × 6 board nailed to the studs will adequately support the hangers. Nail this board across two or three stud space, notching the studs so the board face will be flush with them. The standard height for a lavatory and the center of this board is 31". A medicine cabinet is practically always centered over the

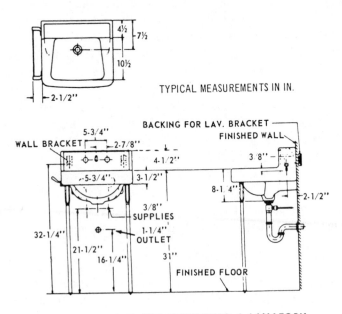

PREPARATIONS FOR INSTALLING A LAVATORY

**Figure 8-10**

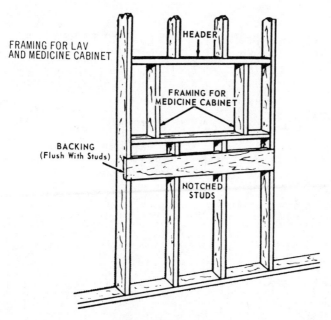

**Figure 8-10 (cont.)**

lavatory. Prepare the framing for this, as in Figure 8–10, to the dimensions required by the cabinet. The cabinet and lavatory will be installed after the wall and floor are finished.

### For a bathtub

The tub end having the waste and pipe connections should be against an inner partition (preferably backed by a closet), so there will be room to make the connections—then to cover them with a removable access panel (which, if in a closet, won't be objectionable). See Figure 8–11 for illustration on how to prepare for installing a tub, and for installing a floor-mounted toilet.

A tub is set down on the sub-flooring and up against the studs, so that the floor and walls can be finished around it to build it in. A typical tub framing is illustrated. The 12 × 6″ notch cut in the sub-flooring is for the branch drain. The access opening should be 3″ high to allow the room for supply pipes. If a shower head is to be installed, provide a 1 × 4″ board to support the riser pipes, set flush into the studs about 5′ above the floor. The tub should be installed when the framing is ready, and before the piping is installed.

A floor-mounted toilet is free-standing and requires no wall brackets. It does require a hole in the floor, under the bowl, for the closet bend

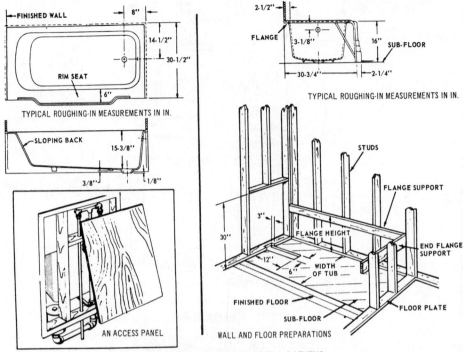

PREPARATIONS FOR INSTALLING A BATHTUB

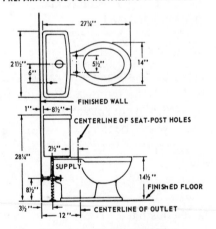

PREPARATIONS FOR INSTALLING
A FLOOR-MOUNTED TOILET

**Figure 8-11**

and flange. When planning this hole be sure to measure out from where the surface of the finished wall will be (not from the studs). A wall-hung toilet requires no hole in the floor, but must be mounted on a special bracket that is secured to the main stack, inside the partitions, at a proper height to support the toilet. This bracket must be attached to the stack prior to finishing the wall.

Whichever type of toilet is used it is not installed until after the walls and floor are finished. See Figure 8–12, for wall-hung toilet illustration.

Most sinks are installed in a cabinet or counter top and require no other support. It is necessary at this time only to locate the supply and drain pipes. The sink will be installed after the cabinet or counter top is made ready.

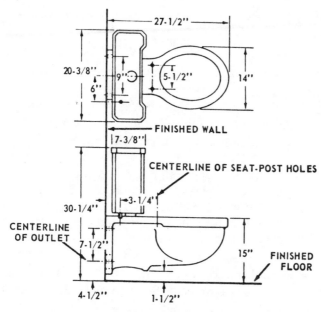

**Figure 8-12**

## INSTALLING A STACK

For any cast-iron main stack, the following specifically pertains to hub-and-spigot type cast-iron pipe. However, if you will disregard all references to hub-and-spigots—and keep in mind the simpler method of coupling pipe lengths and fittings together—these same instructions apply equally well to the no-hub cast-iron pipe.

If the previously installed house-drain wye is more than 6' below the first tee (such as the one for a toilet branch) needed in the stack, begin by permanently installing one 5' pipe vertically in the upright wye hub. Should distance be less than 6', begin by making measurements for cutting off a length of pipe to place this first tee at correct height. See

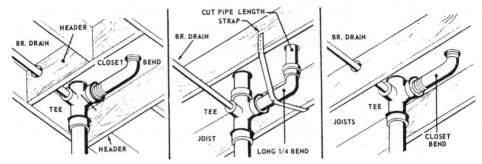

**THREE WAYS OF INSTALLING TEE, TOILET BRANCH AND ONE OTHER BRANCH DRAIN**

**Figure 8-13**

Figure 8–13 for three ways of installing the tee, toilet branch and one branch drain.

When the first tee will be a sanitary tee with the side tap (to hold a closet bend and connect to the tub and/or the lavatory branch drain), there are two heights at which it can be installed.

1. If the sanitary bend will run crosswise to the floor joists in an already prepared header space, the tee will be at a height which will place the top of the closet bend at or slightly above the level of the finished floor in the bathroom above. In this case the tub and lavatory branch drain will run between the joists (to right or left, depending upon whether a RH or LH tapped tee is used).

2. If the closet bend will run with the joists, then the branch drain must run crosswise to them. See Figure 8–13 for details.

   a. You can install the tee and closet bend as seen, then bore holes in all the joists to run the branch drain through them, if drilling them is permissable, as in Figure 8–1.

   b. Or, you may be able to use a ¼ bend at the tee tap to elevate the branch drain above the joists into a partition space.

   c. But if neither of these solutions is practical, you will have to install the tee low enough to run the branch drain beneath the joists. If this is the case you will substitute, in place of the closet bend, a long ¼ bend with a short (spigot at each end) piece of pipe long enough to reach the toilet above. Note that you also can plug the tee tapping, then use as a tapped tee at a higher level for branch drains, to run in the partition space.

Whichever solution you adopt, measure down from the level of the finished bathroom floor to the installed stack pipe or wye to determine the

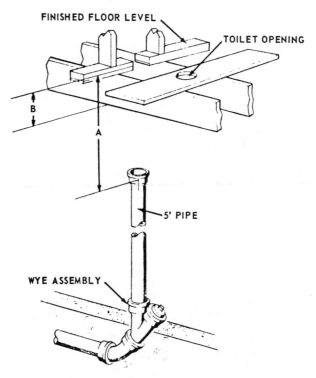

**Figure 8-14**

length of this portion of the run. See Figure 8–14. Also measure from where the center of the stack will be straight across on the floor above to the center of the prepared toilet opening. Using these measurements lay your pipes and fittings out on the floor as shown to mark the pipes for cutting. Remember, either a long ¼ bend or a closet bend can be shortened at the top end. When installed, the closet-bend flange should rest on the finished bathroom floor, encircling the top of the closet bend, or the pipe substituted for it. As in Figure 8–15, all the pieces can be pre-assembled, if desired, then be assembled in the stack. After assembly in the stack the closet bend (or ¼ bend) should be supported by a pipe strap or wood brace beneath its outer end.

Continue to assemble the stack upward. Wherever another tee will be within a full pipe length above, use a similar procedure to measure the tee's location, mark and cut the pipe to place it at the correct height in the stack. If there is a second-floor bath, build the stack up to the proper height, then proceed exactly as for the first-floor bath to install the tee and closet bend. As you build the stack up, don't forget any vent-line tees that must be included.

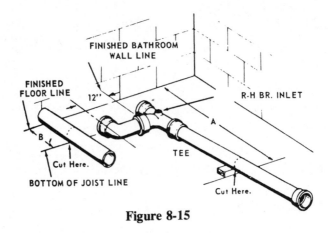

FINISHED BATHROOM
WALL LINE

FINISHED
FLOOR LINE    12″

R-H BR. INLET

A

B

Cut Here.

TEE

BOTTOM OF JOIST LINE

Cut Here.

**Figure 8-15**

*Note*: Be sure to aim any tapped tee tapping correctly. It will help to screw a 6 to 12″ pipe nipple loosely into the tee tapping.

Figure 8–14 shows how to locate a roof opening for the stack, and how to attach the flashing to the roof.

Should a joist, at any point, interfere with the vertical rise of the stack, the stack can be offset by using two ⅛ bends, as shown. Wherever it is necessary to locate an opening above for the stack, use a plumb bob, as shown for locating the roof opening (Figure 8–16). Allow the top of the stack to extend at least 12″ above the roof top. Install the flashing around the stack, under the roof shingles, as in Figure 8–16. Tap it snugly around the pipe and seal the joint with asphalt roofing cement.

### ANY STEEL SECONDARY STACK

Like a main stack, any secondary stack also will rise from a wye, a wye which, before now, has been installed at the end of an underground branch from the house drain, or which is part of an exposed house-drain run. Installation of a secondary stack therefore differs from installation of a main stack only in these respects: 1) Steel pipe (1½″ or 2″ size) rather than cast-iron, generally is used. See Figure 8–17 for a typical steel secondary stack. 2) There will be no toilet branch to install. 3) If, as usually is the case, the stack serves a sink located on an outside wall, the stack will rise through a chase left in the wall for this purpose, and will have to be offset inward from the wall to go through the roof. 4) The final length of pipe that goes out through the roof will usually be a vent increaser—required by code in any area where frost would be likely to block the opening of a smaller (1½″ or 2″ diameter) pipe. 5) This stack need not vent through the roof, and can revent into a main stack, instead.

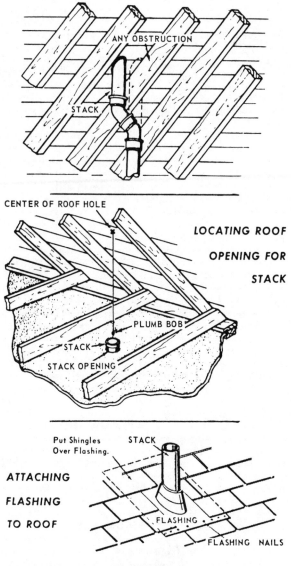

**Figure 8-16**

If the stack does extend up through the roof use the same steel pipe throughout. Measure from the foot-of-stack wye (to be exact, from the shoulder in the hub of this wye) upward to where the first sanitary tee must be located, and cut and thread the first pipe length accordingly, allowing for a caulking spigot at the bottom end. Join the tee and spigot to this length and set the spigot end into the wye hub, bracing the pipe to stand upright if necessary. If another tee above the first one is required,

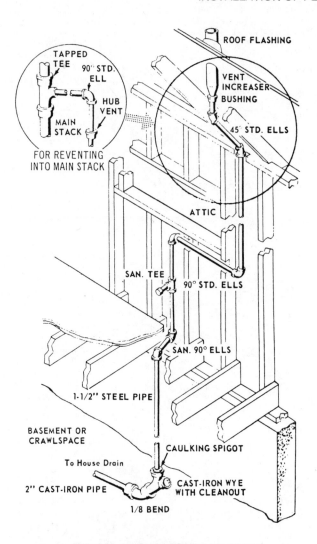

**A TYPICAL STEEL SECONDARY STACK**

**Figure 8-17**

similarly measure the next pipe length, then cut and thread it, join the tee to it, and join these pieces to the first assembly. Continue in this manner up to just below the roof. Check to be certain the tee or tees face correctly (best to screw a nipple loosely into the tee branch for this check), then caulk the bottom end of the run tightly into the wye hub.

Wherever it is possible (as in an attic) the top of a stack that is in outer wall should be offset inward so as to go out through the roof somewhat back from the roof plate (rather than through the roof plate).

This is particularly desirable if a vent increaser is to be used. Use two 45° ells and a short length of pipe between them, assembled to the stack in turn. If no vent increaser is to be used simply top the stack with a final short pipe length to project 12″ or more out through the roof, or install a vent increaser, instead. In either case, finish the roof with flashing, as for the main stack.

If the stack will not extend through the roof, but will run horizontally between floor or in an attic to join the main stack instead, the vent portion always is 1½″ pipe. Since it is impossible to thread a length of pipe or a run of threaded pipe into stationary fittings at each end, there has to be a hub-vent fitting (used like, but instead of a union) somewhere in this vent portion. It is always in a vertical part of the run. Therefore, the easiest procedure is to first complete the waste portion of the run, as preceding, up to (and including) the highest sanitary fitting. Then install a short length of pipe (topped with the hub-vent fitting) in this fitting. Next, start at the vent fitting in the main stack and install the horizontal portion of this vent run, ending with an ell that is in vertical alignment with the hub-vent. Finally, install the remaining vertical length of pipe needed in the ell, then caulk the bottom end of this into the hub-vent fitting, as in Figure 8–17. *Note*: A hub-top tee can sometimes be used to advantage, in place of the hub-vent fitting. Remember that the horizontal vent portion is to pitch slightly upward to the main stack.

## ANY COPPER OR PLASTIC STACK

Figures 8–18 and 19 show these conversions. The procedure for installing stacks of these materials is the same as for cast-iron and/or steel, with these exceptions: Copper and plastic pipes are furnished in 10′ lengths, so the span from the wye at the foot of the stack up to the first tee should be measured and cut in one piece. Copper fittings correspond to

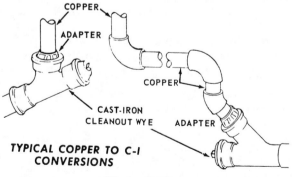

**Figure 8-18**

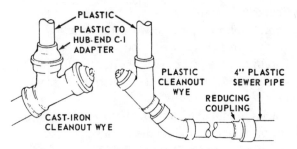

TYPICAL PLASTIC TO C-I OR TO 4-IN. PLASTIC
CONVERSIONS

**Figure 8-19**

cast-iron fittings, so you have the same possible set-up variations for the bathroom as shown for a cast-iron stack. Also, copper fittings can be used in the same manner as steel-pipe sanitary fittings for drainage, or like standard steel-pipe fittings for venting, if reversed.

Plastic fittings are also much the same, except that there is no basic package tee to serve the toilet branch and also (through a side tap) serve another branch to the tub and/or lavatory. When assembling a plastic basic stack, the branches to the tub, lavatory and sink must be joined to the stack through a separate tee or tees above and/or below the toilet-branch tee. This main stack tee can, however, be made to serve all three (tub, lavatory and sink) if the other tees are installed in the branch drain so that it, in turn, will branch off to the three fixtures. Whether you install the tub, etc., branch tee above or below the toilet-branch tee depends entirely upon your house construction. If installed above, the branch will have to run through a partition; if installed below, it will have to run below the joists of the floor above.

## INSTALLING BRANCH DRAINS AND VENT RUNS

Figure 8–20 shows typical waste and vent runs with steel pipe. After a stack is completed with all tees installed and "pointing" in the proper directions, installation of the branch-drain and vent runs is simply a matter of measuring, cutting and installing the required pipe lengths and fittings. Keep these factors in mind:

1. Any fitting through which water will flow must be of the sanitary type, and all horizontal water runs must pitch down at 1″ per 4 to 5′. Such runs can be located in partitions, below joists in a basement or crawlspace, or if there is room in a floor between

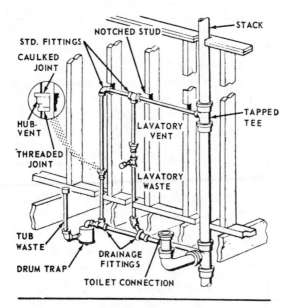

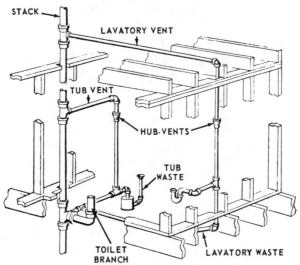

*TYPICAL WASTE AND VENT RUNS — STEEL PIPE*

**Figure 8-20**

joists or across joists if notching of the joists is practical, as in Figure 8–1.

2. Fittings through which gases only will pass can be standard types or, if sanitary fittings are used as in the case of copper, they

should be reversed so the flow is upward instead of downward. All horizontal vent runs should pitch up at 1″ per 5′ or more. Such runs can be located in partitions, above joists in an attic or, like drainage runs, in a floor.

3. If a vent run is to be used it must be taken off from the branch drain it serves at the highest point in the drain. Even if the branch drain serves two or more fixtures, one vent will suffice if properly joined at the highest point (unless your code requires separate reventing of each fixture). If a tee or wye is used the fitting must be the sanitary type, properly installed for the waste flow, and not the gas flow.

4. Two or more branch drains can be joined by sanitary tees or wyes before entering the stack as a single branch so long as all parts slope down, and so long as the common branch is properly sized.

5. Two or more vent runs can be joined by standard tees or wyes before entering the stack, so long as all parts slope up.

6. For details regarding lengths of waste runs, where they may be connected to a stack or house drain, and reventing requirements refer to Chapter 7 and Figures 7–2 and 3.

Figure 8–21 shows how a vent run can be vented directly out through the roof, instead of being revented to the main stack. When venting this way, use a vent increaser.

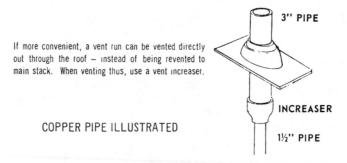

If more convenient, a vent run can be vented directly out through the roof — instead of being revented to main stack. When venting thus, use a vent increaser.

COPPER PIPE ILLUSTRATED

3″ PIPE

INCREASER

1½″ PIPE

**Figure 8-21**

To assemble any drainage run, start at the stack and work outward to the fixture, remembering to install a tee or wye (sanitary) at any point where another branch drain or a vent run must take off. If the branch drain goes to a lavatory, sink or laundry tub, run it through the partition or floor to the point where your roughing-in mark indicates it should be. At this point install a 90° ell. If using copper or plastic, install a pipe length in the ell long enough to project outward beyond the finished wall or floor

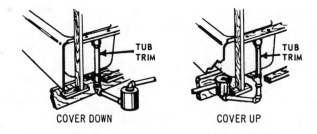

**BATHTUB DRUM-TRAP ARRANGEMENT**

**Figure 8-22**

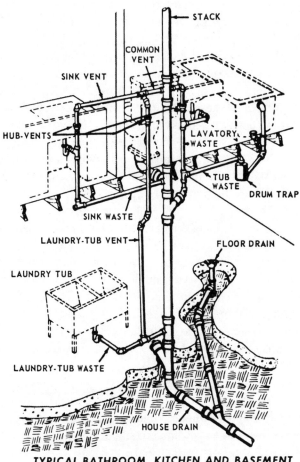

**TYPICAL BATHROOM, KITCHEN AND BASEMENT
FIXTURES WASTE AND VENT RUNS IN STEEL**

**Figure 8-23**

surface by at least 1″ (any excess can easily be sawed off later). If using steel, select any nipple that will accomplish the same and screw it loosely into the ell (a correct length nipple can be permanently installed later). Figure 8–22 shows a bathtub drum-trap arrangement. Figure 8–23 shows typical bath and other fixtures waste and vent runs in steel.

A bathtub normally is installed at the time of roughing-in (See Figure 8–11). The drum trap used is part of and furnished with the drainage system, and is not a fixture trim (Figure 8–22). There are no exposed drainage fittings. Therefore, branch-drain connections to a tub can be completed now. Typical connections are seen in Figure 8–23. If the trap can be reached for service from the basement or crawlspace, install it cover down. If it can be reached through an access panel behind the tub, install it in this area cover up. When there is no other choice, install it in the bathroom floor, with cover up and flush with the finished floor.

To assemble any vent run you can start either at the stack or at the tee where it takes off from a branch drain. In the case of copper or plastic, from whichever end you start simply work on through to the other end to complete the run. In the case of steel, however, remember that you can't join a threaded pipe at two ends simultaneously. Therefore, an unthreaded connection is required to bring the two ends together. This unthreaded connection is a hub-vent fitting—always installed at some vertical portion of the vent run. The bottom part threads onto the pipe below; the pipe above is caulked into the hub at the top of this fitting.

Figure 8–24 shows how to install new branch tees in an existing main stack. If you are adding a fixture for which no provision was made when the stack was built, either an existing stack or an existing house drain must be broken into to effect the branch drain connection. This means that a section of the stack or drain must be broken out and removed. In the case of cast-iron remove the entire length of any piece of pipe broken into; with copper or plastic, saw squarely across the run to remove no more than necessary. In any case, do not leave the stack above a removed portion unsupported—brace it securely beforehand to prevent sagging or falling.

In the case of cast iron the new tee or wye (for the new branch) is installed—and the stack or drain is made whole again by using a sission joint and a short length of pipe or pipes, as required. See Chapter 6.

Figure 8–24 shows a typical tee installation in a stack. Note that the sission-joint stem is cut off so that when the joint is lifted up to join the piece above it, this stem will project no more than necessary into the pipe below. The bottom joint (around the sission-joint stem) should be the last

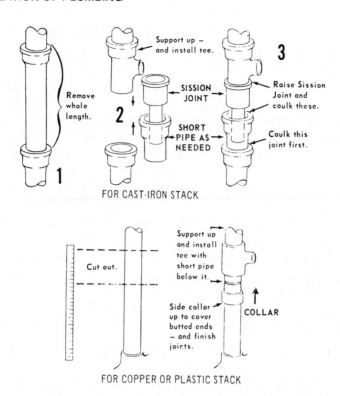

**INSTALLING NEW BRANCH TEE IN AN EXISTING MAIN STACK**

**Figure 8-24**

joint to be caulked—and the pieces above should be supported up until this is done.

In the case of copper or plastic, a collar (to fit snugly around both pipe ends) is needed to complete the last (bottom) joint. To make a collar saw a coupling in half, saw the hub end off a fitting, or saw the top off a pipe cap. Slip this collar down over the end of the pipe at the bottom, then assemble the new pieces to exactly fill the space in the stack, with the two ends that form the bottom joint tightly butted together so that the collar can be slipped up to cover the crack.

## TESTING THE DRAINAGE SYSTEM

If all the joints have been professionally made there should be no cause to fear leakage in your system, which will not be subjected to any severe pressure. However, your code may require testing. The test must be made before installing the house sewer.

To make a test it is necessary to cap or plug every drain opening throughout the system, except the top openings of the stacks. For this reason the test must be made prior to toilet installation and, preferably, prior to installation of any fixtures. The stub pipes, where fixtures are to be installed, can more easily be capped; if a fixture such as a tub or lavatory is installed, the drain opening into the fixture must be plugged.

To cap a steel pipe simply screw on a pipe cap. For copper pipe, solder on a pipe cap to be unsoldered or sawed off later. For plastic pipe wrap cloth around a bend-end wire to be used for pulling the plug out later and plug the inside, leaving about 1″ at the end to be filled with a stiff putty. For cast-iron pipe, do the same, using cloth or crumpled paper. See Figure 8–25. Then fill the last ½″ with a stiff cement mix. To plug a floor drain or a fixture drain use the method suggested for plastic pipe. At the end of the house drain, in the concrete plug used here, you can install a drain cock and hose if desired to drain off the test water when finished.

CAUTION

Do not fill system during freezing weather.

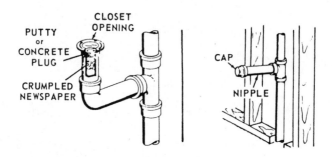

**Figure 8-25**

After plugging all openings fill the system with water to the tops of the stacks. Systematically check all joints throughout the entire system for leakage. When satisfied, drain the system, remove all plugs or caps, and complete the installation. Do not fill the system during freezing weather.

### THE BUILDING SERVICE LINE

If your water source is a public or commercial water service, the water authorites will probably want to install or at least approve your building service line, including a meter if used. This pipe should be below the frost line, deep enough so that traffic won't disturb it, not at any point directly beneath any sanitary sewer pipe and of ample size insofar as allowed to supply all the water you will need at peak load. It can, depending upon the code, be of type K copper tubing or galvanized steel pipe, but the copper is recommended. See Figure 8–26 for a diagram of the meter in a basement.

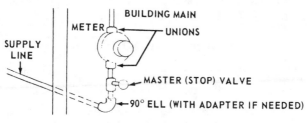

**METER IN BASEMENT**

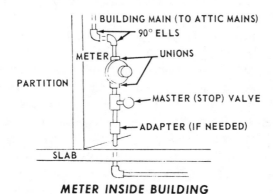

**METER INSIDE BUILDING**

**Figure 8-26**

When there is a basement the meter, if used, is customarily in the basement at the point where the pipe enters. There should be a master valve just ahead of the meter. The meter should be in a location where it can easily be read, but where it will not interfere with your use of the basement. If you have installed an in-wall length of pipe to connect the service line to the outside you will have established the meter location. Install a 90° ell at the point where the pipe comes through the wall, followed by a stop valve, then have the meter installed against the wall following the valve.

Where there is no basement the meter if used is usually in the house, in the kitchen utility room or a hallway, unless local weather permits it to be outside. Whether inside or out there should be a master valve ahead of it. When outside, provision may be required to protect it from vandalism with some type of enclosure, built onto the side of the building. When it is inside, any portion of the pipe exposed to freezing (as in a crawlspace) may have to be enclosed and insulated.

Even though a master valve is used as recommended, some codes or water authorities require a shut-off valve installed in the service line as close as possible to the water-supply main. If so, the water authority should install this.

### THE BUILDING AND OTHER MAINS

An electric water heater can be located wherever it is convenient to the piping and to the electrical connections. A fuel-burning heater, however, must be located near enough to a chimney or outside wall for proper venting of the combustion gases. A water softener or conditioner can be located wherever it will be convenient to service it and to make the piping connections to it.

Usually, there is a single pipe run (see Figure 8–27 for ways of connecting a water softener or conditioner) in the building main from the end of the service line to the heater, where this single run branches, as at a tee, with one branch into the heater and the other branch becoming the cold-water main. If, however, the heater is at some distance from the service line—and if the pipe can be saved by doing so—the building main can be branched right at the end of the service line (or anywhere else) with one branch to the heater and the other becoming the cold-water main. Up to the tee (wherever located) this building main should be the same size pipe (¾" or 1") as the service line. Following the tee the pipe size of each branch should be ¾". The pipe that comes out of the heater, which is the hot water main, also should be ¾". Since the building main carries cold water, any cold-water branch that is convenient to join directly to this may be so joined. For instance, if the main passes by a basement laundry tub, or a point that is convenient to start a cold-water riser to a floor above

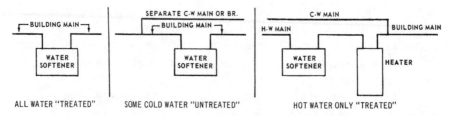

WAYS OF CONNECTING A WATER SOFTENER OR CONDITIONER

**Figure 8-27**

or a run out to a sill cock, then use a tee and join the tub cold-water branch, the cold-water riser or the cold-water sill-cock run to the building main. See Figure 8–28 for a typical water-supply piping in the basement of a one-story house.

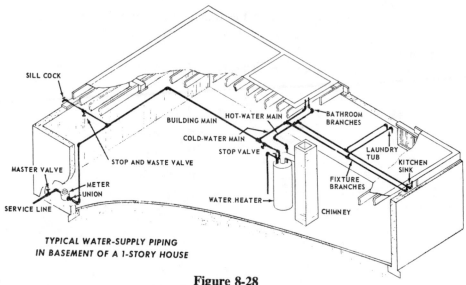

TYPICAL WATER-SUPPLY PIPING
IN BASEMENT OF A 1-STORY HOUSE

**Figure 8-28**

*Note*: For the convenience and appearance it generally is considered preferable to run hot and cold-water mains side-by-side to serve the (also side-by-side) branches to all fixtures requiring hot and cold water. Wherever pipe can be saved by doing otherwise, however, don't hesitate to break this rule.

If all of the water (cold and hot) to be used on the premises, inside and outside, is to be softened or conditioned, the softener or conditioner is installed in the building main ahead of any branch. If all the water, cold and hot, to be used inside only—not the water used for lawn sprinkling, etc.—is to be so treated, then the softener or conditioner is located in the building main ahead of all branches except one which serves the outside. This branch which serves the outside may be a single cold-water branch to one sill cock, or it may be a separate (untreated) cold-water main that will serve two or more sill cocks and/or a sprinkler system, etc. When it is a separate cold-water main it, like other mains, should be ¾" pipe. If desired, it also can be used to serve some inside fixtures should you, for instance, prefer to have unsoftened cold water for drinking.

If only the hot water is to be treated, the softener or conditioner will be installed in the cold-water main, just ahead of the hot-water heater.

Any main (building, cold-water or hot-water) connected to a water heater, water softener or water conditioner should be connected by a union (located as close to the unit as possible). There also should be a stop valve, ahead of the union, in the line through which water flows to the unit. These arrangements allow the unit to be disconnected from the system for servicing, easily and without having to shut down the remainder of the system.

See Figure 8–29 for typical water-system drainage arrangements.

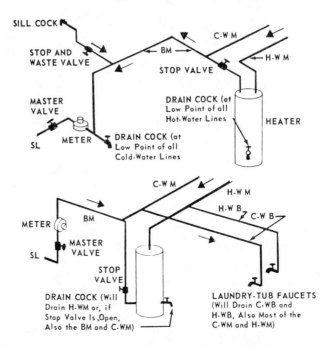

**TYPICAL WATER-SYSTEM DRAINAGE ARRANGEMENTS**

**Figure 8-29**

When installing mains in a basement or crawlspace some provision should be made for draining the system in case repairs are needed, and after testing it. The inlet union connection to your water heater, or the heater drain cock, may serve the purpose. Similarly, the faucets of a basement laundry tub may be made to serve. However, if all or part of the system won't drain to such a place, then you should install a tee and drain the cock at the low point or at each low point, if there are two or more.

This is particularly important if the house is one which may be closed up, unheated, during freezing weather.

If the mains are located in an attic, the low point of the building main will probably be where it joins the service line. Each downward run from the attic will have its own low point—probably the faucets of the fixtures it serves—or the union connection to or the drain cock of the heater, etc., it is connected to.

Figure 8-30 shows typical uses of air chambers. Any water system collects air, and this is dispelled with the water through the faucets. However, if it is excessive, it will cause the faucets to sputter, and the pipes to make a knocking sound when the faucets are turned on. One way to overcome these disadvantages is to provide air chambers where the air in the system will collect, to be bled off in a regulated manner when the faucets are open. An air chamber is a vertical length of pipe 12″ long or more, joined at a tee and capped at the top.

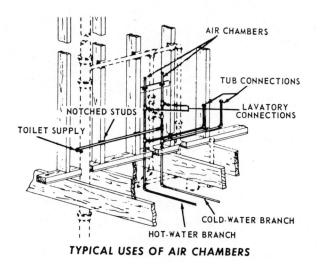

**TYPICAL USES OF AIR CHAMBERS**

**Figure 8-30**

## REMAINDER OF BUILDING PIPING

Whatever the arrangement of the cold and hot-water mains in the basement, crawlspace or attic, it is preferable to run them side-by-side wherever they must pass through partitions, walls or floors. This also applies to the hot and cold-water branches to a fixture. Running them side by side, especially if you are remodeling, will save installation labor. Studs, joists, floor plates, etc., can be more conveniently drilled or notched. Side by side means 4 to 6″ apart—not closer—or the hot-water one will radiate heat to the cold-water one.

Any pipe that serves two or more fixtures or faucets should be considered a main, and be ¾″ size. A pipe that serves just one fixture or faucet is a branch, and usually is ½″ size, though one serving a sill cock may be ¾″ size of the cock will be used to supply several hoses or a sprinkling system.

The valve recommended for installation just ahead of the heater will shut off all hot water (only) in the building. The valve at the end of the supply line will shut off all water, cold and hot. If you want better and more selective control than this, additional stop valves must be used. Selective control is desirable. It avoids having to shut down all or a large part of your water system every time a faucet washer needs replacement or a leak you can't immediately repair develops. See Figure 8–31 for recommended stop valves if no valves are in the fixture supply lines.

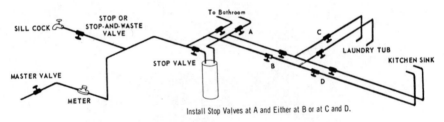

Install Stop Valves at A and Either at B or at C and D.

**RECOMMENDED STOP VALVES IF NO VALVES IN FIXTURE SUPPLY LINES**

**Figure 8-31**

The most selective control is afforded by purchasing fixture supply lines with valves. The hot and/or cold water to each fixture can be shut off alone. Next selective is to install stop valves in the two mains arranged so that the hot or cold water to one or another room can be shut off without affecting other rooms. The least selective arrangement recommended is to have just one extra valve to shut off all outdoor (cold) water alone. Wherever freezing weather occurs this is necessary. In fact, if there is more than one run to the outside, each run should be protected by a stop-and-waste valve, installed where it cannot be frozen and at a low point where the entire outdoor portion of the run can be drained by opening the waste fitting on the valve.

## INSTALLATION PROCEDURES

If threaded pipe is used it is necessary—to avoid use of unwanted unions—to begin each run where it originates at the end of a supply line or from a tee, the heater or softener, etc., and work out to its end. With copper pipe any sequence that seems desirable can be used.

See Figure 8–32 for types of pipe insulation, ways of creating air chambers, and a typical basement main and branch.

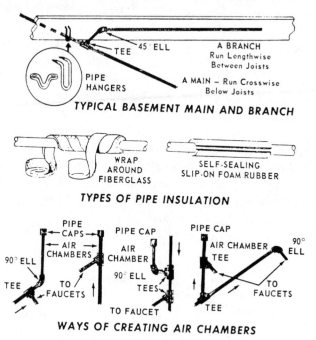

**Figure 8-32**

In a basement or crawlspace pipes usually are run below the joists, then fastened to the joists for support with pipe hangers. They also may be run up between joist, in which case a 45° ell (rather than a 90° ell) is preferred to direct a run from below or above to between the joists. In an attic the pipes usually are laid across the joists (hangers aren't needed) or down between them. If the pipes in a crawlspace or attic are subject to freezing they must be insulated. Also, it may be advisable at least to insulate hot-water pipes, to prevent heat loss. When the pipes are between joists it is much easier to insulate them than when they are below or above. If they are between, insulation blankets can be stretched between the joists to cover them. If they are below or above, the best method is to wrap them with strips of insulation.

Between floor, if pipes cannot be run between joists they should be laid across on top of them, in small-as-possible notches prepared to recieve them. To run them upward or downward into a partition the plaster boards can be drilled or notched, and for crossing through a partition the studs also can be drilled or notched. Wherever pipes must run in an outside wall a cove, in the inner face of the wall, should be provided (when the wall is erected) for them. If there is danger of freezing, the space between a pipe and its cover should be filled with adequate insulation.

A branch run that terminates at a sill cock should project out just far enough to install the cock flush against the wall. Any other branch to a fixture should be run to the point in the wall or floor indicated by your roughing-in mark for the fixture. See Figure 8–33 for ways of installing branches in-wall and in-floor. At this point install a 90° ell and a short pipe or nipple to project out about 1″ beyond the finished surface of the wall or floor. In the case of copper, be generous—any excess projection can be sawed off. In the case of steel, another nipple can later be substituted if necessary. Cap all projecting pipe ends until after the walls and the floors are finished—to prevent plaster, etc., from getting inside. Capping is necessary also for testing.

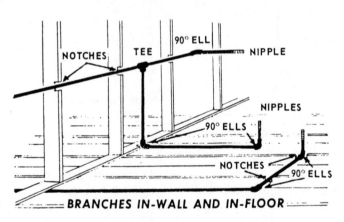

**Figure 8-33**

Figure 8–34 shows an in-wall bathtub connection. *Note*: If the branch run is to have an air chamber as previously recommended, install it in one of the ways illustrated (See Figure 8–32). If a tub is to have in-wall faucets and spout (with or without a shower head also in-wall), the faucet assembly is installed inside the partition behind the tub, with nipples projecting out from the partition to attach the spout and the shower head to, and with openings in the partition to insert the valve stems into their valves.

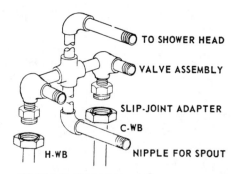

**IN-WALL BATHTUB CONNECTIONS**

**Figure 8-34**

Keep in mind when installing branch runs to a fixture: the cold-water faucet is ordinarily the one at the right side as you face the fixture, and the hot water faucet is at the left side.

Figure 8–35 shows the swing joints or flexible connectors which help you align the connections.

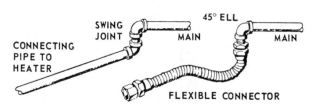

**SWING JOINTS OR FLEXIBLE CONNECTORS HELP YOU ALIGN CONNECTIONS**

**Figure 8-35**

### CONNECTING APPLIANCES

Each appliance, water heater, water softener or conditioner, automatic laundry washer etc., will have fittings and requirements which differ from those of other makes and/or models. To assure making connections properly always refer to the owner's manual furnished with each appli-

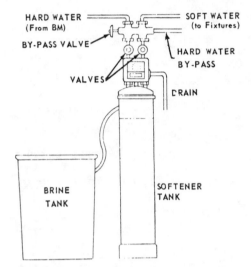

**TYPICAL WATER-SOFTENER CONNECTIONS**

**Figure 8-36**

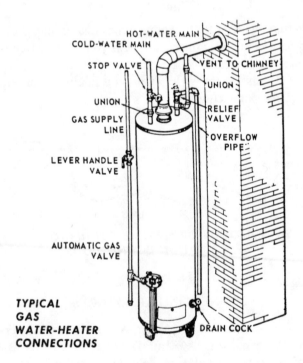

**TYPICAL
GAS
WATER-HEATER
CONNECTIONS**

**Figure 8-37**

ance. Typical connections are shown in Figures 8–36 and 37, these for Wards units.

When installing any appliance keep in mind that it should be as level as possible—also, unions should be used in making pipe connections to it. If gas connections are also to be made, be sure to follow instructions carefully, and to provide proper venting.

If you have difficulty aligning a pipe with the appliance it is to be connected to, a swing joint (Figure 8–35) can be used to facilitate the alignment. Another solution is to use flexible tubing.

## TESTING SYSTEM

A water system should be tested prior to closing the walls and floors. Cap or plug or close with a sill cock all openings in the system except those at the high points (air chambers) or the branch lines. Open all valves, opening the master valve last and just enough to allow water to flow slowly into the system. When water in the system rises to the top of the lowest air chambers, cap them. As water continues to rise, cap each succeeding air chamber that is filled until the last one is capped (no openings left anywhere in the system) and the system is filled with water.

At this time pressure will build up to maximum in the system, whether master valve is fully opened or not. Leave the system under pressure for several hours, then carefully check every joint. If a leak is found, refer to the section on plumbing repairs. After checking, close the master valve and drain the entire system.

## INSTALLING FIXTURES

*Note*: With the exception of the bathtub, fixtures are set after all concealed pipe is completed and the walls are finished. Fixture drainage and water-supply connections are then made to the in-wall or in-floor nipples or stubs. Remove fixtures from their crates carefully to avoid chipping, scratching or denting. When opening crates do not pry against the fixtures or strike them with a hammer. Fixtures can also be damaged by dropping or sliding on a bare tile or concrete floor.

*To install a bathtub*: if not already done at the time drainage piping was installed, attach the chrome-plated tub drain and overflow fittings to their respective drainage-pipe ends. These are threaded to screw into the special ells of the brass drainage fitting connected to the drainage system when the tub was installed. Form a ring of stiff putty around the outside of each fitting behind the flange—to seal it to the tub surface when tightening, unless rubber washers are furnished.

Check the spout and shower-head nipples. Each should project approximately ½". If too long, substitute shorter nipples or, in case of copper pipe, cut off then install steel-pipe male adapters at the ends. Screw the spout and shower head each onto the end of its nipple to butt flush against the wall surface. Screw each valve stem finger tight into its faucet, then tighten the packing nut, which is around the stem, into the faucet threaded boss. Place escutcheon over each stem and secure flush against the wall by tightening the setscrew (use pliers and don't mar it). Attach cold-water handle to the right-hand stem and hot-water handle to the left one. See Figure 8–38 for a typical tub-waste trim procedure.

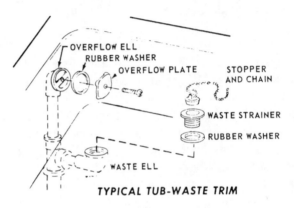

**TYPICAL TUB-WASTE TRIM**

**Figure 8-38**

### A LAVATORY

Attaching faucets is more easily accomplished with the lavatory on the floor, prior to setting it in place. Read the instructions furnished with your faucets, and attach them accordingly. To avoid any seepage of water from around the faucets, spread some stiff putty under each one before tightening down in place.

Figure 8–39 shows how to measure and position the bracket when setting in place. If this is a wall-hung type, position the bracket on the backside of the lavatory to find the distance ("A" in Figure 8–39). Using a carpenter's level, make a horizontal line on the wall to show where the top of the bracket will be, then also mark a vertical line at the centerline of the drainage nipple. Also mark the vertical centerline on the bracket and the back of the fixture. Place the bracket in position against the wall as indicated by the two drawn lines, and mark the bracket screw hole on the wall. Drill the hole in the wall of a proper size for each screw, and install

### Setting in Place

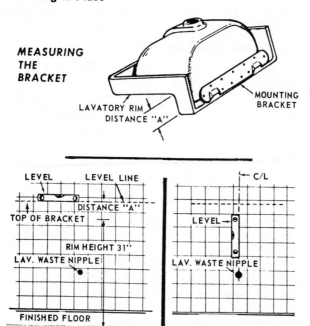

POSITIONING THE BRACKET

**Figure 8-39**

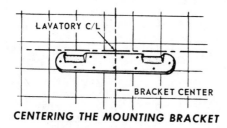

CENTERING THE MOUNTING BRACKET

**Figure 8-40**

the bracket on the wall. See Figure 8–40 for details of centering the mounting bracket.

*Note*: Wood screws are most commonly used. Each should be long enough to penetrate the backing board (inside partition) by 1″. If no backing board and wall is sufficiently solid, use an expanding screw or a toggle bolt.

If the lavatory is unsupported by legs, simply hang it on the bracket, centering it per the centerline marks on it and the bracket. Make certain it is firmly on the bracket before releasing your hold on it. If there are legs, attach these loosely to the lavatory before hanging it, then adjust them to just touch the floor and tighten the adjusting nuts.

A pedestal type is bolted to the floor. Center it in front of the drainage nipple centerline with its back to the wall. Mark the bolt-hole locations in the floor. Drill bolt holes and install the toilet-bowl type bolts that are used in the floor. Set lavatory back in position over the bolts, and secure it to the bolts with the nuts and washer furnished.

The installation of a cabinet type depends entirely upon the cabinet construction.

## CONNECTING THE WASTE

The usual waste fitting furnished with a lavatory is made in two parts: a tailpiece, which hangs from the underside, and the drain, which screws down into the tailpiece from above. There may be a rubber washer to fit under the drain flange and seal it against the basin, and there is usually a brass washer and a nut to tighten the drain in place from the underside before screwing the tailpiece onto it. Or, there may be a rubber washer for the bottom side. If there is no top rubber washer, use putty as suggested for installing a bathtub drain. See Figure 8–41 for a diagram on how to connect a lavatory to waste connections.

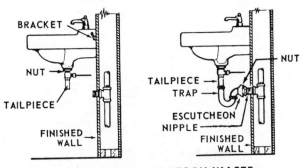

**CONNECTING LAVATORY WASTE**

**Figure 8-41**

*Note*: Take care, when tightening chromed or brass drainage pipes and fittings not to mar them and also not to crush the pipe. Don't use a pipe wrench; use large pliers with a cloth or tape wrapped around the pipe to protect the finish, or on fittings use a monkey wrench or open-end wrench of the proper size.

Check the in-wall or in-floor drainage nipple. It should extend approximately 1″, and the end should be threaded for the trap slip nut (1½″). If necessary, shorten the nipple and install a male adapter at its end. Place the escutcheon over the trap end, then secure the end loosely into the nipple using the slip nut and washer provided with the trap. Fit the other end to the tailpiece and also secure it loosely with a slip nut and washer. Check to be certain that the trap is squared at both ends, then tighten both slip nuts.

## CONNECTING THE WATER SUPPLY

A lavatory supply line may be purchased with or without a stop valve. The valve provides, at nominal cost, the convenience of being able to shut off the water to this one faucet should repair be needed, without affecting the rest of the system. The usual size for a supply pipe is ⅜″ but some are ½″. Also, there usually is a ½″ slip nut fitting at each end, though some may be threaded at the supply end only, instead. See Figure 8–42 for a typical supply-line connection.

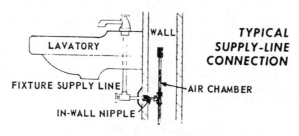

**Figure 8-42**

In any event, the first step is to provide a nipple to project from the finished surface of the wall or floor approximately ½″ to 1″, and to have a proper fitting at its end, of the right size, to receive either the slip nut or the threaded supply-line end as the case may be. This may require a coupling or a ½ × ⅜″ reducing coupling, if the nipple is steel, or a male or female threaded adapter or female threaded reducing adapter, if the nipple is copper.

With the escutcheon in place, on its supply end attach this end of the supply line to the nipple, then use stiff putty to hold the escutcheon against the wall or floor around the joint. Bend the supply line as required to mate its other end with the faucet, and connect it with the slip nut provided.

## A TOILET: CONNECTING THE WASTE

A wall-hung toilet is supported by a closet carrier installed on the stack at the time of the stack assembly, and empties directly into the stack tee through a special fitting furnished with the toilet. Make the installation in accordance with instructions furnished in the owner's manual furnished with the toilet.

A floor-mounted toilet rests on the floor over the end of the branch drain, and is held in place by bolts. See Figure 8–43. Whether the branch drain is a closet bend or an assembly consisting of a ¼ bend and short pipe, the in-floor end must be either flush at the top with the finished floor surface or slightly (not over ½″) below flush. A flange, the rim of which rests on the floor, fits down around the branch-drain end, and must be tightly sealed to it. The rim of the opening in the bottom of the toilet rests on the top of this flange—to seal this joint, which is tightened when the toilet is bolted down. The bolt heads are held under slotted openings in

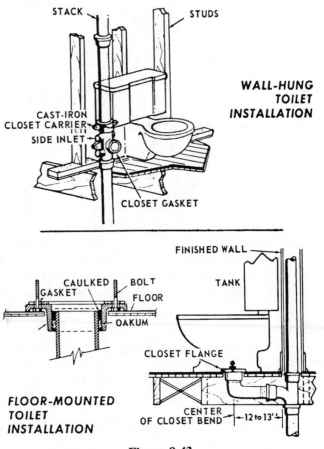

**Figure 8-43**

the flange rim, or in some cases, toilet bowl bolts may be used with their wood-screw ends turned down into the wood floor.

If the branch drain is cast-iron or plastic a cast-iron flange is used. To join it to a cast-iron drain, pack the joint with oakum, fill to the top with molten lead, and caulk the lead in the same manner as for other cast-iron pipe joints. To join it to a plastic drain, pack with oakum then fill to the top with solvent cement. If the branch drain is copper, a copper flange is used and is sweated onto the drain end with solder. In all cases, the flange must rest flat on the floor, with its top level.

When the flange is in place, insert the bolt heads through the flange slots and position them (the bolts) upright so that the toilet can be fitted over them in a position that will align it squarely with the wall behind it. Install the gasket on top of the flange over the bolt ends (or, if putty is to be used, provide a generous ring of it to encircle the opening in the top of the flange—without getting any into the opening). Lower the toilet over the bolt ends, check to make certain it is square with the wall behind it, then use the nuts and washers provided to bolt it firmly down to squeeze the gasket (or squeeze out any excess putty). Attach the toilet tank to the basin back (unless yours is a one-piece toilet) per instructions furnished with the toilet. See Figure 8-44 for two ways of connecting a toilet tank.

When connecting the water supply, the cold-water supply line to the toilet tank is installed in the manner explained for installing lavatory supply lines. When connecting the supply-line end to the projecting end

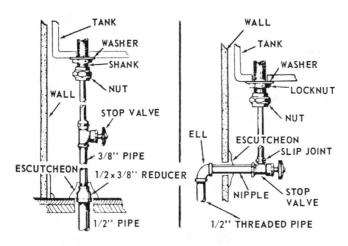

*TWO WAYS OF CONNECTING TOILET TANK*

**Figure 8-44**

of the tank ball-cock stem, hold the ball cock with pliers to prevent it from turning.

Figure 8–45 shows the manner of connecting the waste to a kitchen sink and how to measure the length of the nipple needed.

## Connecting the Waste

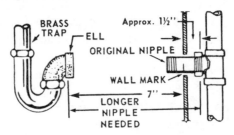

*MEASURING LENGTH OF NIPPLE NEEDED*

**Figure 8-45**

If a sink should be wall hung, it is installed and waste connections are made the same as for a lavatory. Most sinks, however, are supported in cabinets for which they are designed. In such cases the installation of the sink in its cabinet must be made per instructions furnished with it. Before setting the cabinet in place make certain that any holes provided in the bottom and/or back of the cabinet (for passage of drainage and/or water-supply lines into the space below the sink) will align with your respective in-wall or in-floor nipples. If they do not, make new holes as needed. Prior to installing the sink in the cabinet, attach the faucets to it.

Since all connecting piping will be hidden within the cabinet, neatness of appearance is unimportant. Brass and, if needed, steel-pipe fittings are frequently used, and if the drainage nipple or nipples do not exactly line up with the sink drain or drains, whatever fittings needed to align them can be employed. See Figure 8–46 for typical sink waste connections.

In short, if you can make the waste connections in the same manner described for a lavatory, do so; but if necessary, let the in-wall or floor nipple project a bit farther and/or use a drainage ell at its end to obtain the right angle for joining the trap to it.

Several typical waste connections are illustrated. Note that when a garbage disposer is installed in one basin or a two-basin sink it should have a separate trap and waste connection to the branch drain. If it is connected to share the other basin trap its action is likely to force waste up into the other basin.

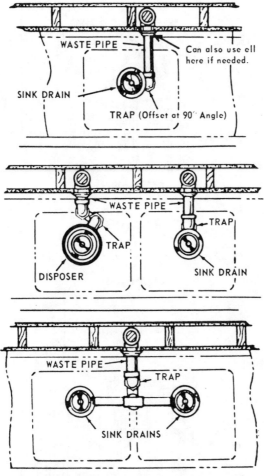

**TYPICAL SINK WASTE CONNECTIONS**
**Figure 8-46**

If the sink is in an island cabinet the drainage run or runs, if there is a disposer requiring a separate connection, and the vent run must go through the bottom of the cabinet and across under the floor to the branch drain for secondary stack in the wall. A typical arrangement is seen in Figure 8–48, with Figure 8–47 showing typical sink water-supply connections. Figure 8–48 also shows laundry tub connections and a clothes washer drain set-up.

### CONNECTING THE WATER SUPPLY

If the sink is in a cabinet, the in-wall or in-floor nipples must be long enough to extend to at least ½″ into the cabinet interior. If desirable (as

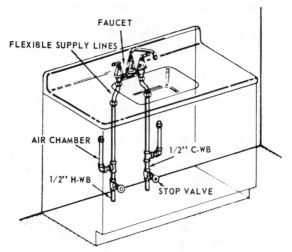

**TYPICAL SINK WATER-SUPPLY CONNECTIONS**

**Figure 8-47**

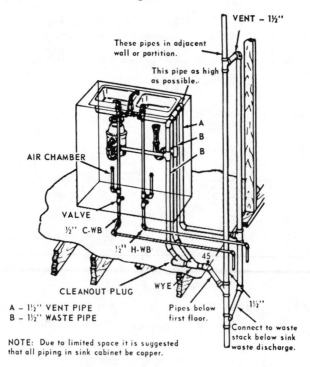

A – 1½" VENT PIPE
B – 1½" WASTE PIPE

NOTE: Due to limited space it is suggested
that all piping in sink cabinet be copper.

**AN ISLAND SINK INSTALLATION**

**Figure 8-48**

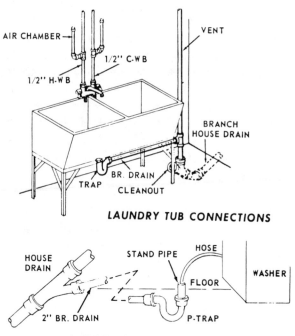

**LAUNDRY TUB CONNECTIONS**

**A CLOTHES WASHER DRAIN**

**Figure 8-48 (cont.)**

would be the case with a wall-hung sink) the nipples may be joined to their faucets by using the same type of supply lines (with or without valves) used for a lavatory, but ½" flexible supply line with slip joints at each end to join this pipe to the faucet. If assembly is of copper, a male threaded adapter is needed to join each flexible line.

## A LAUNDRY TUB

Steel, copper or plastic drainage pipe and fittings, including a P-trap, are commonly used for a laundry tub. The drain fitting inside the tub is threaded to screw onto the top end of a threaded P-trap. If you are using copper or plastic, a male-threaded adapter is needed at this point. Be sure to pitch downward any horizontal run required to join the trap to the drainage system, and use as few bends in this run as possible. One typical set-up is seen in Figure 8–48.

The steel or copper water-supply branch piping is generally extended all the way to the faucets, without use of any flexible supply-line pipe. If desired, air chambers can be built into these lines as illustrated. Each connection to the faucet is made with a slip nut and washer, furnished with the faucet.

## APPLIANCES

Dishwashers and automatic clothes washers are always furnished with specific instructions for making the waste and water-supply connections to them. If a permanent (not a hose-type) waste discharge is installed for either appliance, it should be a separate run connected to the branch drain behind (not ahead of) the trap of any fixture—with a trap of its own. A dishwasher, unless portable, requires a branch drain with an air gap (see Figure 7–8) to prevent backflows from the sink waste. Clothes washers generally have drain hoses. If a more controlled discharge is desired, install a branch drain ending with a vertical pipe rising 12 to 18″ out of the floor, then drop the hose end a foot or so down into this stand pipe. Water-supply connections to these appliances may be made by running the branch lines to them and by using either flexible supply-line pipes—with or without stop valves—or by using hoses with adapters.

# 9

# HOUSE SEWER AND
# FINAL DISPOSAL
# INSTALLATION

## TYPES OF PIPES AND THEIR ASSEMBLY

Cast-iron pipe and fittings—the same as used for a building drainage system (see Chapter 6) may be used for a house sewer—from the house drain to the public sewer or to a septic tank. It also may be used for a gutter drain; but it is never used for a foundation drain or for a disposal-field piping. For assembly refer to Chapter 6, again.

Vitrified-tile pipe is similar in appearance to cast-iron pipe, and is a durable underground pipe that is not deteriorated by contact with soil chemicals. It is somewhat porous and will permit seepage of water through it. For this reason it cannot be used for a house sewer if the line will pass within 10 feet of a well, and should not be used if any tree or shrub with water-seeking roots is close enough to wrap roots around it. It is somewhat fragile, and also should not be used where it will be subjected to strain, as where the earth may settle or under a driveway. It is excellent for use in a foundation drain, in a disposal field and, if conditions permit, for gutter or areaway drainage to a storm sewer or lower ground.

See Figure 9–1 for typical tile pipe and fittings, typical fiber pipe and fittings, and typical private disposal systems.

Underground plastic pipe is available in the same design pipe and fittings used for building drainage, and is assembled in the same manner when root-proof joints are required. Or, it can be loosely assembled

195

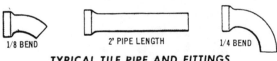

**TYPICAL TILE PIPE AND FITTINGS**

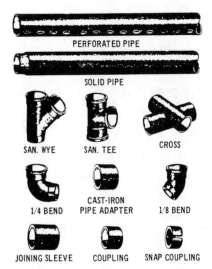

**TYPICAL FIBER PIPE AND FITTINGS**

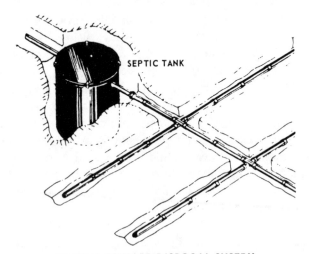

**TYPICAL PRIVATE DISPOSAL SYSTEM**

**Figure 9-1**

where, as in a disposal field, there is no need for root or leakproof joints. The solid pipe is excellent for irrigation. Solid pipe and fittings are available in 10' lengths in 3 and 4" sizes; perforated pipes in 10' lengths in 4" size.

Fiber pipe, like plastic, is available with fittings in both solid and perforated types, and these are used in the same ways mentioned for the plastic pipes. Firm, root, and leakproof joints are quickly and easily made.

See Figure 9–1 for typical tile and fiber pipe and fittings, and a typical private disposal system.

### SEPTIC TANKS

Most septic tanks are made either of steel or concrete. Steel is favored because it is lighter and easier to install. Wards, for example, furnishes steel tanks in 300, 500 and 750-gal. sizes to meet any home owner's need. These tanks are made of heavy-guage welded sheetmetal coated inside and out with acid and rust-resistant asphaltum, for long trouble-free service. Any two or more tanks can be used in combination to provide any greater capacity desired.

### INSTALLING A HOUSE SEWER

If you are connecting to a public sewer, many communities require a licensed workman to install the house sewer, through the local code. And connection to the public sewer requires a permit and must be done by the sewer authorities or a bonded contractor. Even where the code allows, it usually is inadvisable for an improperly equipped person to undertake the installation. Most public sewers are quite deep—up to 20' or more below ground level. Hand trenching the house sewer to such depth would be not only difficult but very hazardous since, with most soils, expert shoring is required to prevent cave-ins.

### IF CONNECTING TO A SEPTIC TANK

The house sewer run to a septic tank will rarely be more than 3' deep at any point. There is no hazard in hand trenching to such shallow depth; but there may be a lot of slow, difficult digging to do. In most communities there is someone with proper excavating machinery to speed this portion of the work, and we recommend that you contract for at least this part of the installation work. See Figure 9–2 for preferred types of house sewers.

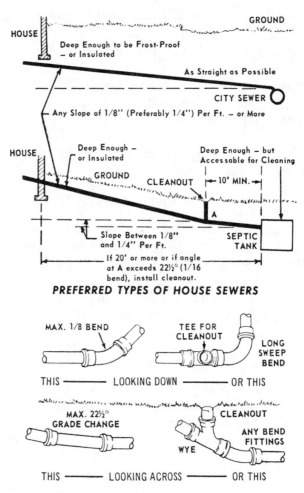

**PREFERRED TYPES OF HOUSE SEWERS**

**Figure 9-2**

## GENERAL REQUIREMENTS

A house sewer should run as straight as possible to final disposal (public sewer or septic tank). That is, unless absolutely necessary, it should not bend, either laterally or vertically.

If any bends are required they should, preferably, not exceed 45° (⅛ bend) in a lateral direction, and not exceed 22½° (¹/₁₆ bend) when changing from a steeper to a shallower grade. If bends in excess of these amounts are required they should be made with long-sweep bends or several smaller bends rather than one short bend.

Whenever it is necessary to bend sewer lines laterally in excess of

45° a cleanout—just ahead of the bend—is required. A cleanout also is required wherever it is necessary to reduce the grade by 22½° or more. A cleanout is provided by installing a wye or a tee with the vertical leg extended upward to ground level and capped—unless the sewer is deeper than 4', in which case a manhole may be required.

The absolute minimum pitch at any point is ⅛" per foot (a drop of 1¼" in 10') and the preferred minimum pitch is ¼" per foot (2½" in 10'). Any pitch in excess of ¼" per foot is permissable. However, to avoid installation of cleanouts, it is better not to vary the pitch unless circumstances make this necessary.

If final disposal is a septic tank the last 10' of sewer ahead of the tank must not slope more than ¼" per foot. A greater slope in this last 10' of sewer will cause water to be dumped into the tank with a violence sufficient to disrupt the chemical liquidation of waste in the tank (and result in the need for too-frequent tank cleanings). Therefore, if the tank is located so much lower than the house that the sewer must pitch sharply down to it, your only choice is to install a cleanout and lessen the sewer grade 10' or more ahead of the tank. In any case, wherever the tank will be more than 20' from the house, it is preferable to have a cleanout 5 to 10' ahead of the tank.

If local code permits, all or parts of a house sewer may be above ground. However, a sewer must be protected against the frost. Hence, it must either be buried deeply enough, or must be properly insulated (with the insulation, if it is exposed, protected from the elements). Also, a septic tank (if used where temperatures drop below freezing) should be deep enough (1 to 2' at the top below ground level) to prevent freezing of water in it—but, preferably, should not be buried too deep (12 to 18" where weather permits) or cleaning it out will become a problem. These factors should be considered when planning the house sewer to a tank.

The house sewer must be a minimum of 4" in diameter, watertight throughout and, preferably (especially if it passes near trees or shrubs) with root-proof joints. If vitrified tile or concrete pipe is used (permitted for connection to a septic tank in some areas), the diameter should be 6". The sewer line should not be any closer than 50' and downhill from any well or other drinking water source. See Figure 9–3 for details as to planning the sewer line measurements and preparing the trench.

## PREPARING THE TRENCH

To dig a trench (or section of a trench), drive stakes into the ground adjacent to one edge of the required trench and at 10' intervals from end

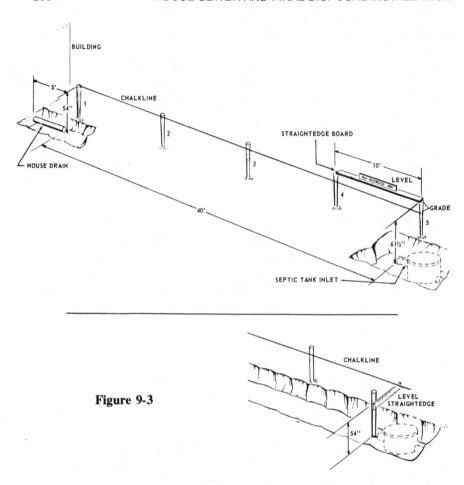

**Figure 9-3**

to end of the run. Using a 10′ straightedge board (like a 1 × 4″ on edge) and a carpenter's level, drive in or add to all of the stakes so that all stake tops are at the same level, a sufficient distance above ground to make the following marks. Start with the stake that is 10′ out from the house drain end and mark it at a measured distance below its top that will indicate the required pitch for the sewer. Similarly, mark all the remaining stakes out to the end of the run. If the pitch is to be 2½″ per 10′, the first stake mark will be 2½″ below the stake top, the next one at 5″ below the stake top, etc. When all the stakes are marked stretch a strong chalkline from one to the next, fastening its starting end on top of the first stake (by building drain end) and tacking it to each other stake at the mark on the stake. This line will now slope down, from the end of the house drain to the far end of the house sewer at the desired pitch for the house sewer.

Start digging your trench to the approximate slope indicated by the chalkline. As soon as practiced, determine the exact vertical distance between the chalkline and the bottom side of the house-drain pipe end. At every point the bottom of your trench must be the same vertical distance below the chalkline above it. To make measurements easy take a straightedge board and mark this distance (from one end) on it. Nail a second straightedge board to it at a right angle, with its top edge touching the mark. Your trench bottom will be properly pitched when you can stand the marked board upright in the trench at any point and the top of the crosspiece will just touch the chalkline (or the mark on any one of the stakes).

Make the trench only as wide as necessary to lay the pipe in it. Also, the trench bottom must be smooth and firmly packed. If necessary to backfill any place where dug too deep, tamp the earth down solidly and true to the grade.

### LAYING THE PIPE AND BACKFILLING

Cast-iron, fiber, underground plastic or (if conditions permit) tile pipe may be used, depending upon local codes and your preference. We recommend fiber or plastic for all-around satisfaction and ease of installation. It is important that the body of each pipe length rest firmly on the trench bottom (see Figure 9–4), to prevent later sagging. Consequently, as pipe is laid, a slight excavation must be made under each hub, coupling or other fitting, as required to permit this (Figure 9–4). When backfilling the trench, start covering the pipe carefully to avoid distrubing it, and tamp fill all around the pipe and each joint to make certain that no air pockets are left to cause later settling. If practical, thorough soaking of the backfill—a layer at a time—will settle the dirt even better than tamping.

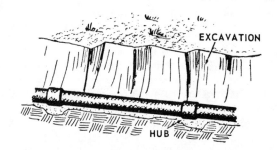

**Figure 9-4**

## INSTALLING A FOUNDATION DRAIN

A foundation drain collects underground water to prevent it from flooding a basement. In some areas such drains are required. In any area where rainfall is average or above, a drain is desirable for a house with a basement. This is particularly true if the house is downgrade from a watershed, or if the ground around the foundation is a fill. The drain is located below the basement floor level, either just outside the foundation or just inside.

See Figure 9–5 for illustration of a drain outside or inside the foundation, and of a foundation drain emptying into a sump.

Perforated fiber or plastic pipe, or tile pipe may be used (if the latter is used, joints are loosely assembled without packing them). All the pipe laid—at one, two, three or four sides of the house—should be graded down to one collection point at a pitch of 1″ per 4′ or better, with a run from this point to some discharge area (a storm sewer or lower ground sloped away from the house). The run from the collection point can be solid pipe with watertight joints, if it is undesirable to have water seep into or out of this run.

Use 4″ pipe throughout. Lay perforated pipe (with holes at the top) or loose-joint pipe in a trench 12 to 18″ wide, and embed the pipe in a gravel or crushed-rock fill as in Figure 9–5. This type fill around the pipe allows only water to be collected in the pipe since it will filter out the earth (and prevent clogging of the pipe). If a solid pipe run is used from the collection point to the disposal, it should be laid in the same manner as a house sewer.

If the local code permits, a foundation drain may empty into the house sewer—joining it at a wye wherever convenient. Or, if you have a suspended house sewer and it is inconvenient to provide a run long enough to join the foundation drain to it underground, you can use a sump pump to lift the foundation drain water into the higher house drain, inside the basement. See Figure 9–5. In such case it is preferable to have the foundation drain inside of the foundation, and to locate the pump in the basement floor. Other branch drains such as one from a laundry tub (never from a toilet) can also be connected to the same sump pump.

## INSTALLING AREAWAY AND GUTTER DRAINS

If the ground slopes away from the house it generally is not desirable to provide a drain for gutter water. A simple concrete slab under the gutter will suffice. However, when water from a gutter, open areaway or patio is

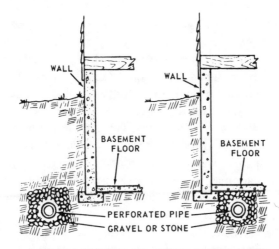

**A DRAIN OUTSIDE OR INSIDE FOUNDATION**

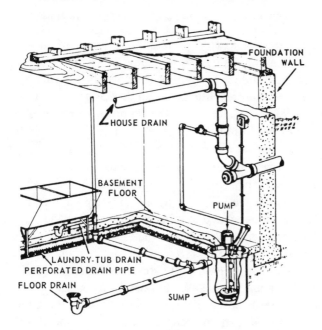

**FOUNDATION DRAIN EMPTYING INTO A SUMP**

**Figure 9-5**

able to collect around the house foundations or where it will be a nuisance, it should be drained off. As with foundation drains, such drains may empty into a sanitary sewer (if the code allows), into a storm sewer, onto downhill ground—or, if there is nowhere else to dispose of the water without damaging your or a neighbor's property, into a seepage pit (as shown following for a private disposal system). Never connect such a drain to a house sewer that ends at a septic tank; but you can connect it to a disposal field beyond the septic tank. Since the object of any drain used as above is to carry water off, cast-iron, tile or solid fiber or plastic pipe should be used—with watertight joints. Perforated pipe or loosely joined tile is used only when (and for that portion of such a drain in which) it is desirable to redistribute water in the ground for irrigation purposes. For instance, gutter water may be carried away from the house in a watertight run, then be allowed to seep back into the soil through perforated pipe laid under the lawn or a flower bed if these are downhill from the house.

Lay solid portions like a house sewer is laid; lay perforated portions in the same way used for a foundation drain in gravel or crushed rock.

### INSTALLING A PRIVATE FINAL DISPOSAL

*Note*: A septic tank with, preferably, a disposal field—or a seepage pit—provides the only satisfactory method of sewage disposal for buildings not served by public sewers. Properly installed, such a system is safe and efficient, and requires little attention. Solids are broken up in the tank by the action of the bacteria and the liquid effluent (resulting fluid) is discharged into the soil at the disposal field or pit. A field is preferable because it provides more ground area for the effluent to seep into and because it affords a larger surface area exposed to sun and wind for completion of effluent disposal by evaporation.

Companies such as Wards supply properly engineered steel tanks with interior baffles designed to minimize turbulence and permit the most effective bacterial action. A concrete tank may be purchased locally or be constructed on the site. Tank size, which is very important, should be sufficient for the tank to hold 100 gallons of water for each person who will live in the house, plus an additional 40 to 50 gallons per person if there is a garbage disposer. For example, the tank for a family of five must hold 700 to 750 gallons if there is a disposer, and 500 if there is no disposer. Determine the number of persons as being the number the house is built to accommodate, not any smaller number who may presently be occupying it. Two or more tanks may be joined together in sequence if necessary.

The tank may be near the house, but should—for health's sake—be at least 100 feet and downhill from any well, cistern, etc. The illustrations (Figures 9–6 and 7) show commonly used types of septic tanks, a septic

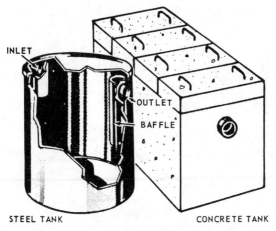

**COMMONLY USED TYPES OF SEPTIC TANKS**

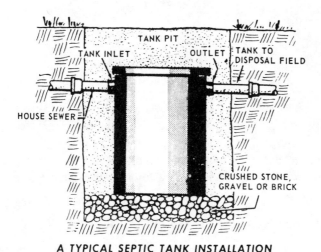

**A TYPICAL SEPTIC TANK INSTALLATION**

**Figure 9-6**

tank installation, and the common measurements used in planning and installing a disposal field. As previously explained, tank depth is determined when laying the house sewer. Make your excavation large enough to conveniently lower the tank in and make the necessary connections, and about 18 to 24″ deeper than the bottom of the tank level. The pit bottom must be firm and reasonably level. Line the bottom with 18 to 24″ of gravel, crushed rock or bricks, and level this accurately before setting the tank on it.

## PLANNING AND INSTALLING A DISPOSAL FIELD

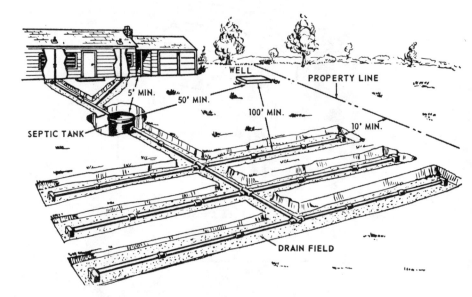

**Figure 9-7**

Make watertight connections between the house and sewer and tank and the tank and the run on out to the disposal field or pit. Afterwards, half fill the tank with water and backfill around it, tamping or soaking the earth in firmly. The water in the tank and immediate backfilling will keep it from being floated out of position by any ground water seepage into the pit.

### PLANNING AND INSTALLING A DISPOSAL FIELD

The ideal location for a disposal field is a flat, large-enough area where there are few trees or shrubs to shade the ground and as far as possible from the house or other occupied buildings. Minimum recommended distances between field and surroundings are seen in Figure 9–7. Check your local code for possible additional requirements.

The field should be flat or nearly so because the seepage piping should be laid to pitch only about 1″ per 10 to 12′, and should not be buried more than 3′ deep or trenching might be a problem. How many feet of pipe the field must contain depends upon the nature of the soil. Accompanying Tables A and B in Figure 9–8 show two methods of determining the pipe footage, while Figure 9–9 shows useful disposal field patterns. Field size is determined by the seepage-pipe footage re-

## TABLE A — PIPE FOOTAGE IN AVER. SOIL

| NATURE OF SOIL | FEET OF PIPE PER PERSON |
|---|---|
| Clean, Coarse Sand or Gravel | 12 |
| Fine Sand or Light Loam | 20 |
| Sand with Some Clay or Loam | 30 |
| Clay with Some Sand or Gravel | 80 |
| Heavy Clay | Unsuitable |

### EXAMPLE

Soil is fine sand with some clay. System normally serves 5 people. Table A shows 30 ft. of drainage pipe is required for each person; five persons will require 5 x 30 = 150 ft. of drainage pipe.

A more accurate method of determining the amount of pipe required is based on a soil "percolation" test which shows the rate of water absorption by the soil. This test is simple and takes only a short time. At the site of the proposed filter bed, simply dig a hole about 1 ft. in diameter to the depth where the disposal piping will be placed. Partially fill the hole with water and, by observation and timing, calculate the amount of pipe required from Table B.

## TABLE B — SOIL PERCOLATION TEST

| TIME REQUIRED FOR WATER TO FALL 1 IN. | FEET OF PIPE PER PERSON |
|---|---|
| 1 Minute | 12 |
| 2 Minutes | 15 |
| 5 Minutes | 20 |
| 10 Minutes | 30 |
| 30 Minutes | 60 |
| 60 Minutes | 80 |
| Over 60 Minutes | Unsuitable |

### EXAMPLE

It takes 30 min. for the water in the percolation test to fall 1 in. Five persons will be served. Table B shows that 60 ft. of drainage pipe will be required for each person; five persons will require 5 x 60 = 300 ft.

**Figure 9-8**

quired, and by these two rules: 1) No run of seepage pipe should exceed 100' in length. 2) Runs should be spaced 10 to 12' apart. Field shape and the pattern in which the seepage-pipe is laid are determined by the slope of the ground and the limitations imposed by boundary lines, buildings, etc. Again, see Figure 9–9.

The run from the tank to the field should be watertight for a distance of at least 5' from the tank. Whether you make the rest of this run

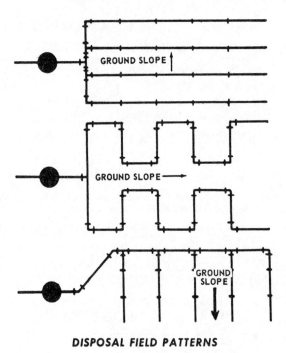

**DISPOSAL FIELD PATTERNS**

**Figure 9-9**

watertight or not depends upon where you want seepage into the soil to begin. All trenching for this run and for the seepage-pipe in the field can be planned and dug to depth in the manner explained for a house sewer. The pitch of the run from tank to field can be 1″ per 5′ or greater, but the seepage-pipe pitch, as mentioned above, should be 1″ per 10 to 12′. All pipe should be buried deep enough to be protected from being crushed by whatever traffic there will be above it.

Perforated fiber or plastic pipe (laid with the holes down) is much the easiest to use as seepage pipe. If tile is used it is assembled with loose joints covered, on top, with building paper or roofing material to keep dirt out. See Figure 9–10 for methods of installing fiber and plastic pipe and laying field tile. If hubless field tile is used the pipes must be laid along the edge of a board buried in the ground beneath them to ensure no sagging, with lengths spaced about ¼″ apart and the joints covered as for tile. If the soil is light and/or sandy, no special backfilling is needed. However, if soil is heavy and non-porous, the trench should be backfilled with gravel or crushed rock to surround the pipe at least 12″ all around.

After laying the field pipe, backfill and level the surface, then sod or

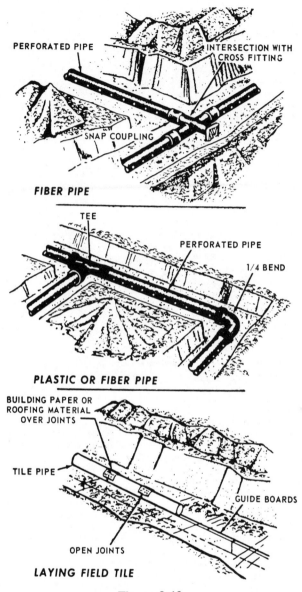

PERFORATED PIPE

INTERSECTION WITH CROSS FITTING

SNAP COUPLING

FIBER PIPE

TEE

PERFORATED PIPE

1/4 BEND

PLASTIC OR FIBER PIPE

BUILDING PAPER OR ROOFING MATERIAL OVER JOINTS

TILE PIPE

GUIDE BOARDS

OPEN JOINTS

LAYING FIELD TILE

**Figure 9-10**

seed it to encourage a thick crop of grass, which will hasten effluent evaporation.

## INSTALLING A SEEPAGE PIT

If a disposal field is not practical and your local code permits, you can substitute a seepage pit (dry well) constructed as illustrated. The pit must have a capacity (below the inlet level) at least as great as the septic tank capapcity. Two or more pits, joined together in sequence, can be used if necessary. If two or more they must be spaced at least 3 times their diameters apart. No pit should be less than 150′ from a wall, etc., less than 20′ from any building, nor less than 10′ from your property line. See Figure 9–11 for details of laying pipe in non-porous soil and typical seepage pit construction.

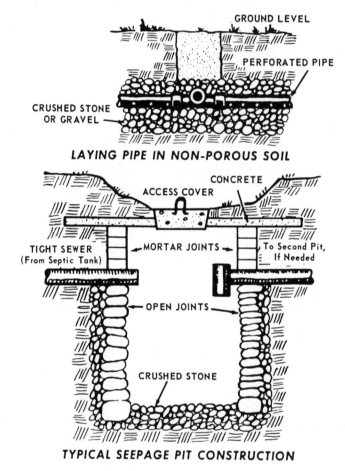

**Figure 9-11**

# 10

# PIPE FITTING TABLES; PRESSURE, FLOW AND THERMAL CHARTS

These charts and tables will prove to be an invaluable resource and reference for you. When you need this information, you will need it quickly, and so we have presented it in tabular and chart form for easy guidance.

Figure 10–1 covers standard pipe data, and can tell you quickly such things as the amount of water a lineal foot of pipe will hold based on its inside and outside diameters, and the weight of that water.

Barlow's Formula, Figure 10–2, is the classic way of finding the relationship between the internal fluid pressure and the stress in the pipe wall, and will give you confidence in your work—and security.

Commercial pipe sizes and wall thicknesses are the stuff of the information in Figure 10–3.

The combination of pipe and water weight per line foot can be found in the numerical chart that is Figure 10–4, while the weight per foot of seamless brass and copper pipe can be found in Figure 10–5.

How much pressure is in the pipe, based on feet head? See Figure 10–6. Figure 10–7 is the chart which shows the feet head of water as compared to the pounds per square inch.

Did you know that the boiling point of water will vary at varying pressures? The full story can be seen in Figure 10–8. Again, you may never need the knowledge contained in these tables, but you can feel secure that if you do, it is here, at your fingertips.

Interested in heat loss statistics? See Figure 10–9, which will give you info on the heat loss from horizontal bare steel pipe, based on the BTU loss per hour per linear foot at 70° room temperature.

Figure 10–10 gives you the total thermal expansion of piping material in inches per 100' above 32° F, in the event you are working (mostly, you shouldn't be) with very close tolerances. But these figures which should be taken into account when calculating your projects.

## STANDARD PIPE DATA

| NOMINAL PIPE DIAM. IN INCHES | ACTUAL INSIDE DIAM. IN INCHES | ACTUAL OUTSIDE DIAM. IN INCHES | WEIGHT PER FOOT POUNDS | LENGTH IN FEET CONTAINING ONE CUBIC FOOT | GALLONS IN ONE LINEAL FOOT |
|---|---|---|---|---|---|
| ⅛ | .269 | .405 | .244 | 2526.000 | .0030 |
| ¼ | .364 | .540 | .424 | 1383.800 | .0054 |
| ⅜ | .493 | .675 | .567 | 754.360 | .0099 |
| ½ | .622 | .840 | .850 | 473.910 | .0158 |
| ¾ | .824 | 1.050 | 1.130 | 270.030 | .0277 |
| 1 | 1.049 | 1.315 | 1.678 | 166.620 | .0449 |
| 1¼ | 1.380 | 1.660 | 2.272 | 96.275 | .0777 |
| 1½ | 1.610 | 1.900 | 2.717 | 70.733 | .1058 |
| 2 | 2.067 | 2.375 | 3.652 | 49.913 | .1743 |
| 2½ | 2.469 | 2.875 | 5.793 | 30.077 | .2487 |
| 3 | 3.068 | 3.500 | 7.575 | 19.479 | .3840 |
| 3½ | 3.548 | 4.000 | 9.109 | 14.565 | .5136 |
| 4 | 4.026 | 4.500 | 10.790 | 11.312 | .6613 |
| 4½ | 4.560 | 5.000 | 12.538 | 9.030 | .8284 |
| 5 | 5.047 | 5.563 | 14.617 | 7.198 | 1.0393 |
| 6 | 6.065 | 6.625 | 18.974 | 4.984 | 1.5008 |
| 8 | 7.981 | 8.625 | 28.554 | 2.878 | 2.5988 |
| 10 | 10.020 | 10.750 | 40.483 | 1.826 | 4.0963 |

**Figure 10-1**

### BARLOW'S FORMULA

Barlow's Formula is a safe, easy method for finding the relationship between internal fluid pressure and stress in the pipe wall. The formula predicts bursting pressures that have been found to be safely within the actual test bursting pressures.

It is interesting to note that the formula uses the "outside diameter" of pipe and is sometimes referred to as the "outside diameter formula."

$$P = \frac{2 \times t \times S}{D}$$

where:

P = internal units pressure, psi
S = unit stress, psi
D = outside diameter of pipe, in.
t = wall thickness, in.

**Figure 10-2**

## COMMERCIAL PIPE SIZES

The following table lists the pipe sizes and wall thicknesses currently established as standard, or specifically:

1. The traditional standard weight, extra strong, and double extra strong pipe.
2. The pipe wall thickness schedules listed in American Standard B36.10, which are applicable to carbon steel.

| NOMINAL PIPE SIZE | OUT-SIDE DIAM. | NOMINAL WALL | | | | | |
|---|---|---|---|---|---|---|---|
| | | SCHED. 5S* | SCHED. 10S* | SCHED. 10 | SCHED. 20 | SCHED. 30 | STAND-ARD† |
| 1/8 | 0.405 | — | 0.049 | — | — | — | 0.068 |
| 1/4 | 0.540 | — | 0.065 | — | — | — | 0.088 |
| 3/8 | 0.675 | — | 0.065 | — | — | — | 0.091 |
| 1/2 | 0.840 | 0.065 | 0.083 | — | — | — | 0.109 |
| 3/4 | 1.050 | 0.065 | 0.083 | — | — | — | 0.113 |
| 1 | 1.315 | 0.065 | 0.109 | — | — | — | 0.133 |
| 1 1/4 | 1.660 | 0.065 | 0.109 | — | — | — | 0.140 |
| 1 1/2 | 1.900 | 0.065 | 0.109 | — | — | — | 0.145 |
| 2 | 2.375 | 0.065 | 0.109 | — | — | — | 0.154 |
| 2 1/2 | 2.875 | 0.083 | 0.120 | — | — | — | 0.203 |
| 3 | 3.500 | 0.083 | 0.120 | — | — | — | 0.216 |
| 3 1/2 | 4.000 | 0.083 | 0.120 | — | — | — | 0.226 |
| 4 | 4.500 | 0.083 | 0.120 | — | — | — | 0.237 |
| 5 | 5.563 | 0.109 | 0.134 | — | — | — | 0.258 |
| 6 | 6.625 | 0.109 | 0.134 | — | — | — | 0.280 |
| 8 | 8.625 | 0.109 | 0.148 | — | 0.250 | 0.277 | 0.322 |
| 10 | 10.750 | 0.134 | 0.165 | — | 0.250 | 0.307 | 0.365 |
| 12 | 12.750 | 0.156 | 0.180 | — | 0.250 | 0.330 | 0.375 |
| 14 O.D. | 14.000 | 0.156 | 0.250 | 0.250 | 0.312 | 0.375 | 0.375 |
| 16 O.D. | 16.000 | 0.165 | 0.250 | 0.250 | 0.312 | 0.375 | 0.375 |
| 18 O.D. | 18.000 | 0.165 | 0.250 | 0.250 | 0.312 | 0.438 | 0.375 |
| 20 O.D. | 20.000 | 0.188 | 0.250 | 0.250 | 0.375 | 0.500 | 0.375 |
| 22 O.D. | 22.000 | 0.188 | 0.250 | 0.250 | 0.375 | 0.500 | 0.375 |
| 24 O.D. | 24.000 | 0.218 | 0.250 | 0.250 | 0.375 | 0.562 | 0.375 |
| 26 O.D. | 26.000 | — | — | 0.312 | 0.500 | — | 0.375 |
| 28 O.D. | 28.000 | — | — | 0.312 | 0.500 | 0.625 | 0.375 |
| 30 O.D. | 30.000 | 0.250 | 0.312 | 0.312 | 0.500 | 0.625 | 0.375 |
| 32 O.D. | 32.000 | — | — | 0.312 | 0.500 | 0.625 | 0.375 |
| 34 O.D. | 34.000 | — | — | 0.312 | 0.500 | 0.625 | 0.375 |
| 36 O.D. | 42.000 | — | — | 0.312 | 0.500 | 0.625 | 0.375 |
| 42 O.D. | 42.000 | — | — | — | 0.375 | — | — |

All dimensions are given in inches.

The decimal thicknesses listed for the respective pipe sizes represent their nominal or average wall dimensions. The actual thicknesses may be as much as 12.5% under the nominal thickness because of mill tolerance. Thicknesses shown in light face for Schedule 60 and heavier pipe are not currently supplied by the mills, unless a certain minimum tonnage is ordered.

**Figure 10-3**

## AND WALL THICKNESSES

3. The pipe wall thickness schedules listed in American Standard B36.19, and ASTM Specification A409, which are applicable *only* to corrosion resistant materials. (NOTE: Schedule 10S is also available in carbon steel in sizes 12″ and smaller.)

**ASA-B36.10 and B36.19**

| THICKNESS FOR | | | | | | | | |
|---|---|---|---|---|---|---|---|---|
| SCHED. 40 | SCHED. 60 | EXTRA STRONG‡ | SCHED. 80 | SCHED. 100 | SCHED. 120 | SCHED. 140 | SCHED. 160 | XX STRONG |
| 0.068 | – | 0.095 | 0.095 | – | – | – | – | – |
| 0.088 | – | 0.119 | 0.119 | – | – | – | – | – |
| 0.091 | – | 0.126 | 0.126 | – | – | – | – | – |
| 0.109 | – | 0.147 | 0.147 | – | – | – | 0.188 | 0.294 |
| 0.113 | – | 0.154 | 0.154 | – | – | – | 0.219 | 0.308 |
| 0.133 | – | 0.179 | 0.179 | – | – | – | 0.250 | 0.358 |
| 0.140 | – | 0.191 | 0.191 | – | – | – | 0.250 | 0.382 |
| 0.145 | – | 0.200 | 0.200 | – | – | – | 0.281 | 0.400 |
| 0.154 | – | 0.218 | 0.218 | – | – | – | 0.344 | 0.436 |
| 0.203 | – | 0.276 | 0.276 | – | – | – | 0.375 | 0.552 |
| 0.216 | – | 0.300 | 0.300 | – | – | – | 0.438 | 0.600 |
| 0.226 | – | 0.318 | 0.318 | – | – | – | – | – |
| 0.237 | – | 0.337 | 0.337 | – | 0.438 | – | 0.531 | 0.674 |
| 0.258 | – | 0.375 | 0.375 | – | 0.500 | – | 0.625 | 0.750 |
| 0.280 | – | 0.432 | 0.432 | – | 0.562 | – | 0.719 | 0.864 |
| 0.322 | 0.406 | 0.500 | 0.500 | 0.594 | 0.719 | 0.812 | 0.906 | 0.875 |
| 0.365 | 0.500 | 0.500 | 0.594 | 0.719 | 0.844 | 1.000 | 1.125 | 1.000 |
| 0.406 | 0.562 | 0.500 | 0.688 | 0.844 | 1.000 | 1.125 | 1.312 | 1.000 |
| 0.438 | 0.594 | 0.500 | 0.750 | 0.938 | 1.094 | 1.250 | 1.406 | – |
| 0.500 | 0.656 | 0.500 | 0.844 | 1.031 | 1.219 | 1.438 | 1.594 | – |
| 0.562 | 0.750 | 0.500 | 0.938 | 1.156 | 1.375 | 1.562 | 1.781 | – |
| 0.594 | 0.812 | 0.500 | 1.031 | 1.281 | 1.500 | 1.750 | 1.969 | – |
| – | 0.875 | 0.500 | 1.125 | 1.375 | 1.625 | 1.875 | 2.125 | – |
| 0.688 | 0.969 | 0.500 | 1.218 | 1.531 | 1.812 | 2.062 | 2.344 | – |
| – | – | 0.500 | – | – | – | – | – | – |
| – | – | 0.500 | – | – | – | – | – | – |
| – | – | 0.500 | – | – | – | – | – | – |
| 0.688 | – | 0.500 | – | – | – | – | – | – |
| 0.688 | – | 0.500 | – | – | – | – | – | – |
| 0.750 | – | 0.500 | – | – | – | – | – | – |
| – | – | 0.500 | – | – | – | – | – | – |

*Schedules 5S and 10S are available in corrosion resistant materials and Schedule 10S is also available in carbon steel.

†Thicknesses shown in italics are available also in stainless steel, under the designation Schedule 40S.

‡Thicknesses shown in italics are available also in stainless steel, under the designation Schedule 80S.

**Figure 10-3 (cont.)**

## PIPE AND WATER WEIGHT PER LINE FOOT

| NOM. PIPE SIZE | WEIGHT OF: | | WEIGHT OF: | |
|---|---|---|---|---|
| | STD. PIPE | WATER | XS PIPE | WATER |
| ½ ⎫ | .851 | .132 | 1.088 | .101 |
| ¾ ⎪ | 1.131 | .231 | 1.474 | .187 |
| 1 ⎬ STD. | 1.679 | .374 | 2.172 | .311 |
| 1¼ ⎪ | 2.273 | .648 | 2.997 | .555 |
| 1½ ⎭ | 2.718 | .882 | 3.632 | .765 |
| 2 ⎫ | 3.653 | 1.453 | 5.022 | 1.278 |
| 2½ ⎪ | 5.794 | 2.073 | 7.662 | 1.835 |
| 3 ⎬ STD. | 7.580 | 3.200 | 10.250 | 2.860 |
| 3½ ⎪ | 9.110 | 4.280 | 12.510 | 3.850 |
| 4 ⎭ | 10.790 | 5.510 | 14.990 | 4.980 |
| 5 | 14.620 | 8.660 | 20.780 | 7.880 |
| 6 | 18.980 | 12.510 | 28.580 | 11.290 |
| 8(.322) | 28.560 | 21.680 | 43.400 | 19.800 |
| 10(.365) | 40.500 | 34.100 | 54.700 | 32.300 |
| 12(.375) | 49.600 | 49.000 | 65.400 | 47.000 |
| 14(.375) | 54.600 | 59.700 | 72.100 | 57.500 |
| 16(.375) | 62.600 | 79.100 | 82.800 | 76.500 |
| 18(.375) | 70.600 | 101.200 | 93.500 | 98.300 |
| 20 | 78.600 | 126.000 | 104.100 | 122.800 |
| 24 | 94.600 | 183.800 | 125.500 | 179.900 |
| 30 | 118.700 | 291.000 | 157.600 | 286.000 |

**Figure 10-4**

## WEIGHT PER FOOT OF
## SEAMLESS BRASS AND COPPER PIPE

| NOMINAL PIPE SIZE | REGULAR | | | EXTRA STRONG | | |
|---|---|---|---|---|---|---|
| | YELLOW BRASS | RED BRASS | COPPER | YELLOW BRASS | RED BRASS | COPPER |
| ½ | 0.91 | 0.93 | 0.96 | 1.19 | 1.23 | 1.25 |
| ¾ | 1.23 | 1.27 | 1.30 | 1.62 | 1.67 | 1.71 |
| 1 | 1.73 | 1.78 | 1.82 | 2.39 | 2.46 | 2.51 |
| 1¼ | 2.56 | 2.63 | 2.69 | 3.29 | 3.39 | 3.46 |
| 1½ | 3.04 | 3.13 | 3.20 | 3.99 | 4.10 | 4.19 |
| 2 | 4.01 | 4.12 | 4.22 | 5.51 | 5.67 | 5.80 |

**Figure 10-5**

## WATER PRESSURE TO FEET HEAD

| POUNDS PER SQUARE INCH | FEET HEAD | POUNDS PER SQUARE INCH | FEET HEAD |
|---|---|---|---|
| 1 | 2.31 | 100 | 230.90 |
| 2 | 4.62 | 110 | 253.93 |
| 3 | 6.93 | 120 | 277.07 |
| 4 | 9.24 | 130 | 300.16 |
| 5 | 11.54 | 140 | 323.25 |
| 6 | 13.85 | 150 | 346.34 |
| 7 | 16.16 | 160 | 369.43 |
| 8 | 18.47 | 170 | 392.52 |
| 9 | 20.78 | 180 | 415.61 |
| 10 | 23.09 | 200 | 461.78 |
| 15 | 34.63 | 250 | 577.24 |
| 20 | 46.18 | 300 | 692.69 |
| 25 | 57.72 | 350 | 808.13 |
| 30 | 69.27 | 400 | 922.58 |
| 40 | 92.36 | 500 | 1154.48 |
| 50 | 115.45 | 600 | 1385.39 |
| 60 | 138.54 | 700 | 1616.30 |
| 70 | 161.63 | 800 | 1847.20 |
| 80 | 184.72 | 900 | 2078.10 |
| 90 | 207.81 | 1000 | 2309.00 |

NOTE: One pound of pressure per square inch of water equals 2.309 feet of water at 62° Fahrenheit. Therefore, to find the feet head of water for any pressure not given in the table above, multiply the pressure pounds per square inch by 2.309.

**Figure 10-6**

## FEET HEAD OF WATER TO PSI

| FEET HEAD | POUNDS PER SQUARE INCH | FEET HEAD | POUNDS PER SQUARE INCH |
|---|---|---|---|
| 1 | .43 | 100 | 43.31 |
| 2 | .87 | 110 | 47.64 |
| 3 | 1.30 | 120 | 51.97 |
| 4 | 1.73 | 130 | 56.30 |
| 5 | 2.17 | 140 | 60.63 |
| 6 | 2.60 | 150 | 64.96 |
| 7 | 3.03 | 160 | 69.29 |
| 8 | 3.46 | 170 | 73.63 |
| 9 | 3.90 | 180 | 77.96 |
| 10 | 4.33 | 200 | 86.62 |
| 15 | 6.50 | 250 | 108.27 |
| 20 | 8.66 | 300 | 129.93 |
| 25 | 10.83 | 350 | 151.58 |
| 30 | 12.99 | 400 | 173.24 |
| 40 | 17.32 | 500 | 216.55 |
| 50 | 21.65 | 600 | 259.85 |
| 60 | 25.99 | 700 | 303.16 |
| 70 | 30.32 | 800 | 346.47 |
| 80 | 34.65 | 900 | 389.78 |
| 90 | 38.98 | 1000 | 433.00 |

NOTE: One foot of water at 62° Fahrenheit equals .433 pound pressure per square inch. To find the pressure per square inch for any feet head not given in the table above, multiply the feet head by .433.

**Figure 10-7**

## BOILING POINTS OF WATER
## AT VARIOUS PRESSURES

| VACUUM, IN INCHES OF MERCURY | BOILING POINT | VACUUM, IN INCHES OF MERCURY | BOILING POINT |
|---|---|---|---|
| 29 | 76.62 | 7 | 198.87 |
| 28 | 99.93 | 6 | 200.96 |
| 27 | 114.22 | 5 | 202.25 |
| 26 | 124.77 | 4 | 204.85 |
| 25 | 133.22 | 3 | 206.70 |
| 24 | 140.31 | 2 | 208.50 |
| 23 | 146.45 | 1 | 210.25 |
| 22 | 151.87 | Gauge Lbs. | |
| 21 | 156.75 | 0 | 212.0 |
| 20 | 161.19 | 1 | 215.6 |
| 19 | 165.24 | 2 | 218.5 |
| 18 | 169.00 | 4 | 224.4 |
| 17 | 172.51 | 6 | 229.8 |
| 16 | 175.80 | 8 | 234.8 |
| 15 | 178.91 | 10 | 239.4 |
| 14 | 181.82 | 15 | 249.8 |
| 13 | 184.61 | 25 | 266.8 |
| 12 | 187.21 | 50 | 297.7 |
| 11 | 189.75 | 75 | 320.1 |
| 10 | 192.19 | 100 | 337.9 |
| 9 | 194.50 | 125 | 352.9 |
| 8 | 196.73 | 200 | 387.9 |

**Figure 10-8**

## HEAT LOSSES FROM HORIZONTAL
## BARE STEEL PIPE

(BTU per hour per linear foot at 70°F room temperature)

| NOM. PIPE SIZE | HOT WATER (180°F) | STEAM 5 PSIG (20 PSIA) |
|---|---|---|
| ½ | 60 | 96 |
| ¾ | 73 | 118 |
| 1 | 90 | 144 |
| 1¼ | 112 | 179 |
| 1½ | 126 | 202 |
| 2 | 155 | 248 |
| 2½ | 185 | 296 |
| 3 | 221 | 355 |
| 3½ | 244 | 401 |
| 4 | 279 | 448 |

**Figure 10-9**

## TOTAL THERMAL EXPANSION OF PIPING
## MATERIAL IN INCHES PER 100 FT. ABOVE 32°F.

| TEMPER-ATURE °F | CARBON AND CARBON MOLY STEEL | CAST IRON | COPPER | BRASS AND BRONZE | WROUGHT IRON |
|---|---|---|---|---|---|
| 32 | 0 | 0 | 0 | 0 | 0 |
| 100 | 0.5 | 0.5 | 0.8 | 0.8 | 0.5 |
| 150 | 0.8 | 0.8 | 1.4 | 1.4 | 0.9 |
| 200 | 1.2 | 1.2 | 2.0 | 2.0 | 1.3 |
| 250 | 1.7 | 1.5 | 2.7 | 2.6 | 1.7 |
| 300 | 2.0 | 1.9 | 3.3 | 3.2 | 2.2 |
| 350 | 2.5 | 2.3 | 4.0 | 3.9 | 2.6 |
| 400 | 2.9 | 2.7 | 4.7 | 4.6 | 3.1 |
| 450 | 3.4 | 3.1 | 5.3 | 5.2 | 3.6 |
| 500 | 3.8 | 3.5 | 6.0 | 5.9 | 4.1 |
| 550 | 4.3 | 3.9 | 6.7 | 6.5 | 4.6 |
| 600 | 4.8 | 4.4 | 7.4 | 7.2 | 5.2 |
| 650 | 5.3 | 4.8 | 8.2 | 7.9 | 5.6 |
| 700 | 5.9 | 5.3 | 9.0 | 8.5 | 6.1 |
| 750 | 6.4 | 5.8 | — | — | 6.7 |
| 800 | 7.0 | 6.3 | — | — | 7.2 |
| 850 | 7.4 | — | — | — | — |
| 900 | 8.0 | — | — | — | — |
| 950 | 8.5 | — | — | — | — |
| 1000 | 9.1 | — | — | — | — |

**Figure 10-10**

# 11

# CONVERSION AND DEFINITION CHARTS AND LISTS

More information on converting information from one type to another is listed in this valuable chapter, as is some referral information on definitions and abbreviations and formulas in common plumbing use. Again, these may not be referred to constantly, but the time may come when this information will be invaluable to you, and we furninsh it for that purpose.

Figure 11–1 tells you the typical BTU values of fuels. Check these values against costs when you have a choice of fuels and want to determine which is the best for you in economic terms.

Figure 11–2 gives you the decimal equivalents of fractions, and Figure 11–3 tells you about minutes converted to decimals of a degree (60 minutes to a degree, 360 degrees in a perfect circle).

Useful definitions of plumbing terms are included as Figure 11–4, while Figure 11–5 is a list of common abbreviations you will run across as an amateur plumber.

Figure 11–7 is helpful when you want to change inches into millimeters, tons into pounds, water pounds into gallons, etc.

Knowledge of the area and volume of pipe and tanks, etc., are important, and when you need that information, you can refer to Figure 11–8. Figure 11–9 offers esoteric information, but it may come in handy if you must drill pipe, know its hardness, and pressure potential.

Now, you have acquired all the information you need to know to begin, and successfully complete, plumbing projects in your home.

## TYPICAL BTU VALUES OF FUELS

**ASTM RANK**

| SOLIDS | BTU VALUES PER POUND |
|---|---|
| Anthracite Class I | 11,230 |
| Bituminous Class II Group 1 | 14,100 |
| Bituminous Class II Group 3 | 13,080 |
| Sub-Bituminous Class III Group 1 | 10,810 |
| Sub-Bituminous Class III Group 2 | 9,670 |

| LIQUIDS | BTU VALUES PER GAL. |
|---|---|
| Fuel Oil No. 1 | 138,870 |
| Fuel Oil No. 2 | 143,390 |
| Fuel Oil No. 4 | 144,130 |
| Fuel Oil No. 5 | 142,720 |
| Fuel Oil No. 6 | 137,275 |

| GASES | BTU VALUES PER CU. FT. |
|---|---|
| Natural Gas | 935 to 1132 |
| Producers Gas | 163 |
| Illuminating Gas | 534 |
| Mixed (Coke oven and water gas) | 545 |

**Figure 11-1**

## DECIMAL EQUIVALENTS OF FRACTIONS

| INCHES | DECIMAL OF AN INCH | INCHES | DECIMAL OF AN INCH |
|---|---|---|---|
| 1/64 | .015625 | 33/64 | .515625 |
| 1/32 | .03125 | 17/32 | .53125 |
| 3/64 | .046875 | 35/64 | .546875 |
| 1/16 | .0625 | 9/16 | .5625 |
| 5/64 | .078125 | 37/64 | .578125 |
| 3/32 | .09375 | 19/32 | .59375 |
| 7/64 | .109375 | 39/64 | .609375 |
| 1/8 | .125 | 5/8 | .625 |
| 9/64 | .140625 | 41/64 | .640625 |
| 5/32 | .15625 | 21/32 | .65625 |
| 11/64 | .171875 | 43/64 | .671875 |
| 3/16 | .1875 | 11/16 | .6875 |
| 13/64 | .203125 | 45/64 | .703125 |
| 7/32 | .21875 | 23/32 | .71875 |
| 15/64 | .234375 | 47/64 | .734375 |
| 1/4 | .25 | 3/4 | .75 |
| 17/64 | .265625 | 49/64 | .765625 |
| 9/32 | .28125 | 25/32 | .78125 |
| 19/64 | .296875 | 51/64 | .796875 |
| 5/16 | .3125 | 13/16 | .8125 |
| 21/64 | .328125 | 53/64 | .828125 |
| 1/3 | .333 | 27/32 | .84375 |
| 11/32 | .34375 | 55/64 | .859375 |
| 23/64 | .359375 | 7/8 | .875 |
| 3/8 | .375 | 57/64 | .890625 |
| 25/64 | .390625 | 29/32 | .90625 |
| 13/32 | .40625 | 59/64 | .921875 |
| 27/64 | .421875 | 15/16 | .9375 |
| 7/16 | .4375 | 61/64 | .953125 |
| 29/64 | .453125 | 31/32 | .96875 |
| 15/32 | .46875 | 63/64 | .984375 |
| 31/64 | .484375 | 1 | 1. |
| 1/2 | .5 | | |

**Figure 11-2**

## MINUTES CONVERTED TO DECIMALS OF A DEGREE

| MIN. | DEG. | MIN. | DEG. | MIN. | DEG. | MIN. | DEG. | MIN. | DEG. | MIN. | DEG. |
|------|------|------|------|------|------|------|------|------|------|------|------|
| 1 | .0166 | 11 | .1833 | 21 | .3500 | 31 | .5166 | 41 | .6833 | 51 | .8500 |
| 2 | .0333 | 12 | .2000 | 22 | .3666 | 32 | .5333 | 42 | .7000 | 52 | .8666 |
| 3 | .0500 | 13 | .2166 | 23 | .3833 | 33 | .5500 | 43 | .7166 | 53 | .8833 |
| 4 | .0666 | 14 | .2333 | 24 | .4000 | 34 | .5666 | 44 | .7333 | 54 | .9000 |
| 5 | .0833 | 15 | .2500 | 25 | .4166 | 35 | .5833 | 45 | .7500 | 55 | .9166 |
| 6 | .1000 | 16 | .2666 | 26 | .4333 | 36 | .6000 | 46 | .7666 | 56 | .9333 |
| 7 | .1166 | 17 | .2833 | 27 | .4500 | 37 | .6166 | 47 | .7833 | 57 | .9500 |
| 8 | .1333 | 18 | .3000 | 28 | .4666 | 38 | .6333 | 48 | .8000 | 58 | .9666 |
| 9 | .1500 | 19 | .3166 | 29 | .4833 | 39 | .6500 | 49 | .8166 | 59 | .9833 |
| 10 | .1666 | 20 | .3333 | 30 | .5000 | 40 | .6666 | 50 | .8333 | 60 | 1.0000 |

**Figure 11-3**

## USEFUL DEFINITIONS

ALLOY STEEL: A steel which owes its distinctive properties to elements other than carbon.

AREA OF A CIRCLE: The measurement of the surface within a circle. To find the area of a circle, multiply the product of the radius times the radius by Pi (3.142). Commonly written $A = \pi r^2$.

BRAZE WELD OR BRAZING: A process of joining metals using a nonferrous filler metal or alloy, the melting point of which is higher than 800°F but lower than that of the metals to be joined.

BUTT WELD: A circumferential weld in pipe fusing the abutting pipe walls completely from inside wall to outside wall.

CARBON STEEL: A steel which owes its distinctive properties chiefly to the various percentages of carbon (as distinguished from the other elements) which it contains.

CIRCUMFERENCE OF A CIRCLE: The measurement around the perimeter of a circle. To find the circumference, multiply Pi (3.142) by the diameter. (Commonly written as $\pi d$).

COEFFICIENT OF EXPANSION: A number indicating the degree of expansion or contraction of a substance.

The coefficient of expansion is not constant and varies with changes in temperature. For linear expansion it is expressed as the change in length of one unit of length of a substance having one degree rise in temperature.

**Figure 11-4**

**DEFINITIONS** (Continued)

CORROSION: The gradual destruction or alteration of a metal or alloy caused by direct chemical attack or by electrochemical reaction.

CREEP: The plastic flow of pipe within a system; the permanent set in metal caused by stresses at high temperatures. Generally associated with a time rate of deformation.

DIAMETER OF A CIRCLE: A straight line drawn through the center of a circle from one extreme edge to the other. Equal to twice the radius.

DUCTILITY: The property of elongation, above the elastic limit, but under the tensile strength.

A measure of ductility is the percentage of elongation of the fractured piece over its original length.

ELASTIC LIMIT: The greatest stress which a material can withstand without a permanent deformation after release of the stress.

EROSION: The gradual destruction of metal or other material by the abrasive action of liquids, gases, solids or mixtures thereof.

RADIUS OF A CIRCLE: A straight line drawn from the center to the extreme edge of a circle.

SOCKET FITTING: A fitting used to join pipe in which the pipe is inserted into the fitting. A fillet weld is then made around the edge of the fitting and the outside wall of the pipe.

SOLDERING: A method of joining metals using fusable alloys, usually tin and lead, having melting points under 700°F.

STRAIN: Change of shape or size of a body produced by the action of a stress.

**Figure 11-4 (cont.)**

**DEFINITIONS** (Continued)

STRESS: The intensity of the internal, distributed forces which resist a change in the form of a body. When external forces act on a body they are resisted by reactions within the body which are termed stresses.

TENSILE STRESS: One that resists a force tending to pull a body apart.

COMPRESSIVE STRESS: One that resists a force tending to crush a body.

SHEARING STRESS: One that resists a force tending to make one layer of a body slide across another layer.

TORSIONAL STRESS: One that resists forces tending to twist a body.

TENSILE STRENGTH: The maximum tensile stress which a material will develop. The tensile strength is usually considered to be the load in pounds per square inch at which a test specimen ruptures.

TURBULENCE: Any deviation from parallel flow in a pipe due to rough inner walls, obstructions or directional changes.

VELOCITY: Time rate of motion in a given direction and sense, usually expressed in feet per second.

VOLUME OF A PIPE: The measurement of the space within the walls of the pipe. To find the volume of a pipe, multiply the length (or height) of the pipe by the product of the inside radius times the inside radius by Pi (3.142). Commonly written as $V = h\pi r^2$.

WELDING: A process of joining metals by heating until they are fused together, or by heating and applying pressure until there is a plastic joining action. Filler metal may or may not be used.

YIELD STRENGTH: The stress at which a material exhibits a specified limiting permanent set.

**Figure 11-4 (cont.)**

## LIST OF ABBREVIATIONS

Abbreviations conform to the practice of the American Standard Abbreviations for Scientific and Engineering Terms, ASA Z10.1.

| | |
|---|---|
| abs | Absolute |
| AGA | American Gas Association |
| AISI | American Iron and Steel Institute |
| Amer Std | American Standard |
| API | American Petroleum Institute |
| ASA | American Standards Association |
| ASHVE | American Society of Heating and Ventilating Engineers |
| ASME | American Society of Mechanical Engineers |
| ASTM | American Society for Testing Materials |
| AWWA | American Water Works Association |
| B & S | Bell and spigot or Brown & Sharpe (gauge) |
| bbl | Barrel |
| Btu | British thermal unit(s) |
| C | Centigrade |
| cfm | Cubic feet per minute |
| cfs | Cubic feet per second |
| CI | Cast iron |
| CS | Cast steel |
| Comp | Companion |
| C to F | Center to face |
| °C | Degrees Centigrade |
| °F | Degrees Fahrenheit |
| diam | Diameter |
| dwg | Drawing |
| ex-hy | Extra-heavy |
| F&D | Faced and drilled |
| F | Fahrenheit |
| F to F | Face to face |
| flg | Flange or flanges |

**Figure 11-5**

| | |
|---|---|
| flgd | Flanged |
| g | Gage or gauge |
| hex | Hexagonal |
| hg | mercury |
| IBBM | Iron body bronze (or brass) mounted |
| ID | Inside diameter |
| kw | Kilowatt(s) |
| MI | Malleable iron |
| max | Maximum |
| min | Minimum |
| mtd | Mounted |
| MSS | Manufacturers Standardization Society (of Valve and Fittings Industry) |
| NEWWA | New England Water Works Association |
| NPS | Nominal pipe size (formerly IPS for iron pipe size) |
| OD | Outside diameter |
| OS&Y | Outside screw and yoke |
| OWG | Oil, water, gas (see WOG) |
| psig | Pounds per square inch, gage |
| red | Reducing |
| sch or sched | Schedule |
| scd | Screwed |
| SF | Semifinished |
| Spec | Specification |
| SSP | Steam service pressure |
| SSU | Seconds Saybolt Universal |
| Std | Standard |
| Trans | Transactions |
| WOG | Water, oil, gas (see OWG) |
| WWP | Working water pressure |
| XS | Extra strong |
| XXS | Double extra strong |

**Figure 11-5 (cont.)**

## UNIT CONVERSIONS
### FLOW

| | |
|---|---|
| 1 gpm | = 0.134 cu. ft. per min. |
| | = 500 lb. per hr. x sp. gr. |
| 500 lb. per hr. | = 1 gpm ÷ sp. gr. |
| 1 cu. ft. per min. (cfm) | = 448.8 gal. per hr. (gph) |

### POWER

| | |
|---|---|
| 1 Btu per hr. | = 0.293 watt |
| | = 12.96 ft. lb. per min. |
| | = 0.00039 hp |
| 1 ton refrigeration (U.S.) | = 288,000 Btu per 24 hr. |
| | = 12,000 Btu per hr. |
| | = 200 Btu per min. |
| | = 83.33 lb. ice melted per hr. from and at 32°F. |
| | = 2000 lb. ice melted per 24 hr. from and at 32°F. |
| 1 hp | = 550 ft. lb. per sec. |
| | = 746 watt |
| | = 2545 Btu per hr. |
| 1 boiler hp | = 33,480 Btu per hr. |
| | = 34.5 lb. water evap. per hr. from and at 212°F. |
| | = 9.8 kw. |
| 1 kw. | = 3413 Btu per hr. |

### MASS

| | |
|---|---|
| 1 lb. (avoir.) | = 16 oz. (avoir.) |
| | = 7000 grain |
| 1 ton (short) | = 2000 lb. |
| 1 ton (long) | = 2240 lb. |

### PRESSURE

| | |
|---|---|
| 1 lb. per sq. in. | = 2.31 ft. water at 60°F |
| | = 2.04 in. hg at 60°F. |
| 1 ft. water at 60°F | = 0.433 lb. per sq. in. |
| | = 0.884 in. hg at 60°F |
| 1 in. Hg at 60°F | = 0.49 lb. per sq. in. |
| | = 1.13 ft. water at 60°F |
| lb. per sq. in. Absolute (psia) | = lb. per sq. in. gauge (psig) + 14.7 |

**Figure 11-6**

## TEMPERATURE

°C      $= (°F - 32) \times 5/9$

## VOLUME

1 gal. (U.S.)      $= 128$ fl. oz. (U.S.)
     $= 231$ cu. in.
     $= 0.833$ gal. (Brit.)

1 cu. ft.      $= 7.48$ gal. (U.S.)

## WEIGHT OF WATER

1 cu. ft. at 50°F. weighs 62.41 lb.
1 gal. at 50°F. weighs 8.34 lb.
1 cu. ft. of ice weighs 57.2 lb.
Water is at its greatest density at 39.2°F.
1 cu. ft. at 39.2°F. weighs 62.43 lb.

## WEIGHT OF LIQUID

1 gal. (U.S.)      $= 8.34$ lb. $\times$ sp. gr.
1 cu. ft.      $= 62.4$ lb. $\times$ sp. gr.

1 lb.      $= 0.12$ U.S. gal. $\div$ sp. gr.
     $= 0.016$ cu. ft. $\div$ sp. gr.

## WORK

1 Btu (mean)      $= 778$ ft. lb.
     $= 0.293$ watt hr.
     $= 1/180$ of heat required to change temp of 1 lb. water from 32°F to 212°F

1 hp-hr      $= 2545$ Btu (mean)
     $= 0.746$ kwhr

1 Kwhr      $= 3413$ Btu (mean)
     $= 1.34$ hp-hr

**Figure 11-6 (cont.)**

## STANDARD CONVERSIONS

| TO CHANGE | TO | MULTIPLY BY |
|---|---|---|
| Inches | Feet | 0.0833 |
| Inches | Millimeters | 25.4 |
| Feet | Inches | 12 |
| Feet | Yards | 0.3333 |
| Yards | Feet | 3 |
| Square inches | Square feet | 0.00694 |
| Square feet | Square inches | 144 |
| Square feet | Square yards | 0.11111 |
| Square yards | Square feet | 9 |
| Cubic inches | Cubic feet | 0.00058 |
| Cubic feet | Cubic inches | 1728 |
| Cubic feet | Cubic yards | 0.03703 |
| Cubic yards | Cubic feet | 27 |
| Cubic inches | Gallons | 0.00433 |
| Cubic feet | Gallons | 7.48 |
| Gallons | Cubic inches | 231 |
| Gallons | Cubic feet | 0.1337 |
| Gallons | Pounds of water | 8.33 |
| Pounds of water | Gallons | 0.12004 |
| Ounces | Pounds | 0.0625 |
| Pounds | Ounces | 16 |
| Inches of water | Pounds per square inch | 0.0361 |
| Inches of water | Inches of mercury | 0.0735 |
| Inches of water | Ounces per square inch | 0.578 |
| Inches of water | Pounds per square foot | 5.2 |
| Inches of mercury | Inches of water | 13.6 |
| Inches of mercury | Feet of water | 1.1333 |
| Inches of mercury | Pounds per square inch | 0.4914 |
| Ounces per square inch | Inches of mercury | 0.127 |
| Ounces per square inch | Inches of water | 1.733 |
| Pounds per square inch | Inches of water | 27.72 |
| Pounds per square inch | Feet of water | 2.310 |
| Pounds per square inch | Inches of mercury | 2.04 |
| Pounds per square inch | Atmospheres | 0.0681 |
| Feet of water | Pounds per square inch | 0.434 |
| Feet of water | Pounds per square foot | 62.5 |
| Feet of water | Inches of mercury | 0.8824 |
| Atmospheres | Pounds per square inch | 14.696 |
| Atmospheres | Inches of mercury | 29.92 |
| Atmospheres | Feet of water | 34 |
| Long tons | Pounds | 2240 |
| Short tons | Pounds | 2000 |
| Short tons | Long tons | 0.89285 |

**Figure 11-7**

## FORMULAS

**Where:**

$A$ = Area; $A_1$ = Surface area of solids;
$V$ = Volume; $C$ = Circumference

### Circle

$A = 3.142 \times R \times R$

$C = 3.142 \times D$

$R = \dfrac{D}{2}$

$D = 2 \times R$

### Ellipse

$A = 3.142 \times A \times B$

$C = 6.283 \times \dfrac{\sqrt{A^2 + B^2}}{2}$

### Parallelogram

$A = H \times L$

### Rectangle

$A = W \times L$

### Sector of circle

$A = \dfrac{3.142 \times R \times R \times \alpha}{360}$

$L = .01745 \times R \times \alpha$

$\alpha = \dfrac{L}{.01745 \times R}$

$R = \dfrac{L}{.01745 \times \alpha}$

### Trapezoid

$A = H \times \dfrac{L_1 + L_2}{2}$

### Triangle

$A = \dfrac{W \times H}{2}$

**Figure 11-8**

### Cone

$A_1 = 3.142 \times R \times S + 3.142 \times R \times R$

$V = 1.047 \times R \times R \times H$

### Cylinder

$A_1 = 6.283 \times R \times R + 6.283 \times R \times R$

$V = 3.142 \times R \times R \times H$

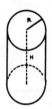

### Elliptical Tanks

$V = 3.142 \times A \times B \times H$

$A_1 = 6.283 \times \dfrac{\sqrt{A^2 + B^2}}{2} \times H + 6.283 \times A \times B$

### Rectangular solid

$A_1 = 2[W \times L + L \times H + H \times W]$

$V = W \times L \times H$

### Sphere

$A_1 = 12.56 \times R \times R$

$V = 4.188 \times R \times R \times R$

For above containers:

Capacity in gallons $= \dfrac{V}{231}$ when V is in cubic inches.

Capacity in gallons $= 7.48 \times V$ when V is in cubic feet.

**Figure 11-8 (cont.)**

## HARDNESS CONVERSION NUMBERS

| BRINELL INDENTATION DIAMETER, MM. | BRINELL HARDNESS NO. — 10-MM. BALL 3000-KG. LOAD, STANDARD OR TUNGSTEN CARBIDE BALL | DIAMOND PYRAMID HARDNESS NUMBER, 50-KG. LOAD | ROCKWELL HARDNESS NUMBER | | ROCKWELL SUPERFICIAL HARDNESS NUMBER SUPERFICIAL BRALE PENETRATOR | | | SHORE SCLEROSCOPE HARDNESS NUMBER | TENSILE STRENGTH (APPROX.) 1000 PSI. |
|---|---|---|---|---|---|---|---|---|---|
| | | | B-SCALE 100-KG. LOAD 1/16 IN. DIA. BALL | C-SCALE 150-KG. LOAD BRALE PENETRATOR | 15-N SCALE 15-KG. LOAD | 30-N SCALE 30-KG. LOAD | 45-N SCALE 45-KG. LOAD | | |
| 2.95 | 429 | 455 | — | 45.7 | 83.4 | 64.6 | 49.9 | 61 | 217 |
| 3.00 | 415 | 440 | —· | 44.5 | 82.8 | 63.5 | 48.4 | 59 | 210 |
| 3.05 | 401 | 425 | -- | 43.1 | 82.0 | 62.3 | 46.9 | 58 | 202 |
| 3.10 | 388 | 410 | -- | 41.8 | 81.4 | 61.1 | 45.3 | 56 | 195 |
| 3.15 | 375 | 396 | — | 40.4 | 80.6 | 59.9 | 43.6 | 54 | 188 |
| 3.20 | 363 | 383 | ·· | 39.1 | 80.0 | 58.7 | 42.0 | 52 | 182 |
| 3.25 | 352 | 372 | (110.0) | 37.9 | 79.3 | 57.6 | 40.5 | 51 | 176 |
| 3.30 | 341 | 360 | (109.0) | 36.9 | 78.6 | 56.4 | 39.1 | 50 | 170 |
| 3.35 | 331 | 350 | (108.5) | 35.5 | 78.0 | 55.4 | 37.8 | 48 | 166 |
| 3.40 | 321 | 339 | (108.0) | 34.3 | 77.3 | 54.3 | 36.4 | 47 | 160 |
| 3.45 | 311 | 328 | (107.5) | 33.1 | 76.7 | 53.3 | 34.4 | 46 | 155 |
| 3.50 | 302 | 319 | (107.0) | 32.1 | 76.1 | 52.2 | 33.8 | 45 | 150 |
| 3.55 | 293 | 309 | (106.0) | 30.9 | 75.5 | 51.2 | 32.4 | 43 | 145 |
| 3.60 | 285 | 301 | (105.5) | 29.9 | 75.0 | 50.3 | 31.2 | — | 141 |
| 3.65 | 277 | 292 | (104.5) | 28.8 | 74.4 | 49.3 | 29.9 | 41 | 137 |
| 3.70 | 269 | 284 | (104.0) | 27.6 | 73.7 | 48.3 | 28.5 | 40 | 133 |
| 3.75 | 262 | 276 | (103.0) | 26.6 | 73.1 | 47.3 | 27.3 | 39 | 129 |
| 3.80 | 255 | 269 | (102.0) | 25.4 | 72.5 | 46.2 | 26.0 | 38 | 126 |
| 3.85 | 248 | 261 | (101.0) | 24.2 | 71.7 | 45.1 | 24.5 | 37 | 122 |
| 3.90 | 241 | 253 | 100.0 | 22.8 | 70.9 | 43.9 | 22.8 | 36 | 118 |
| 3.95 | 235 | 247 | 99.0 | 21.7 | 70.3 | 42.9 | 21.5 | 35 | 115 |
| 4.00 | 229 | 241 | 98.2 | 20.5 | 69.7 | 41.9 | 20.1 | 34 | 111 |
| 4.05 | 223 | 234 | 97.3 | (18.8) | — | — | — | — | — |
| 4.10 | 217 | 228 | 96.4 | (17.5) | — | — | — | 33 | 105 |
| 4.15 | 212 | 222 | 95.5 | (16.0) | — | — | — | — | 102 |
| 4.20 | 207 | 218 | 94.6 | (15.2) | — | — | — | 32 | 100 |
| 4.25 | 201 | 212 | 93.8 | (13.8) | — | — | — | 31 | 98 |
| 4.30 | 197 | 207 | 92.8 | (12.7) | — | — | — | 30 | 95 |
| 4.35 | 192 | 202 | 91.9 | (11.5) | — | — | — | 29 | 93 |
| 4.40 | 187 | 196 | 90.7 | (10.0) | — | — | — | — | 90 |
| 4.45 | 183 | 192 | 90.0 | (9.0) | — | — | — | 28 | 89 |
| 4.50 | 179 | 188 | 89.0 | (8.0) | — | — | — | 27 | 87 |
| 4.55 | 174 | 182 | 87.8 | (6.4) | —· | — | --· | — | 85 |
| 4.60 | 170 | 178 | 86.8 | (5.4) | — | — | — | 26 | 83 |
| 4.65 | 167 | 175 | 86.0 | (4.4) | — | — | — | — | 81 |
| 4.70 | 163 | 171 | 85.0 | (3.3) | ·-· | — | — | 25 | 79 |
| 4.80 | 156 | 163 | 82.9 | (0.9) | — | — | — | — | 76 |
| 4.90 | 149 | 156 | 80.8 | — | — | — | — | 23 | 73 |
| 5.00 | 143 | 150 | 78.7 | — | — | — | — | 22 | 71 |
| 5.10 | 137 | 143 | 76.4 | —· | —· | — | — | 21 | 67 |
| 5.20 | 131 | 137 | 74.0 | — | -·- | — | — | — | 65 |
| 5.30 | 126 | 132 | 72.0 | — | ··· | ··· | — | 20 | 63 |
| 5.40 | 121 | 127 | 69.8 | — | — | — | — | 19 | 60 |
| 5.50 | 116 | 122 | 67.6 | — | — | — | — | 18 | 58 |
| 5.60 | 111 | 117 | 65.7 | — | — | — | — | 15 | 56 |

NOTE: Values in ( ) are beyond normal range; given for information only.

**Figure 11-9**

# INDEX